Organic Chemistry: A Conceptual Approach

Organic Chemistry: A Conceptual Approach

G. H. Williams *Professor of Chemistry* Ph.D., D.Sc., F.R.I.C.
Bedford College, University of London

Heinemann Educational Books
London

Heinemann Educational Books

LONDON EDINBURGH MELBOURNE AUCKLAND TORONTO
HONG KONG SINGAPORE KUALA LUMPUR NEW DELHI
NAIROBI JOHANNESBURG LUSAKA IBADAN KINGSTON

ISBN 0 435 65930 8

First published 1977

Published by Heinemann Educational Books Ltd
48 Charles Street, London W1X 8AH

Printed by William Clowes & Sons Ltd
London, Beccles and Colchester

Preface

The extensive modernization of Advanced Level Syllabuses which has taken place during the last few years has created a need for new textbooks in which not only the material but also the approach to the subject has been brought up to date. In particular, while some texts are available which are closely linked to particular teaching philosophies (such as 'learning by discovery'), there appears to the author to be a need for an Organic Chemistry text at Advanced level in which the material is presented as directly, clearly, and concisely as possible. That is what this book attempts to do while treating the subject in an up-to-date fashion.

The material is organized according to functional groups, and aliphatic and aromatic compounds are not separated. Instead, the properties of functional groups are in each instance related to the framework to which they are attached. Much repetition is thus avoided and a logical approach to organic chemistry is encouraged. Reaction mechanisms are discussed as appropriate, and two reactions, namely nucleophilic substitution at a saturated carbon atom and aromatic nitration, are selected for particularly detailed study. The necessary physical and theoretical basis for such a modern study of Organic Chemistry is laid in the first few chapters. A real effort has been made to limit the size of the book as far as possible; for example, experimental details of preparations etc. are not given since there is no shortage of sources of these. However, the essential factual and theoretical material is all included. Some biochemical topics are introduced in the final chapter. Although some of this biochemical material may be outside some Advanced Level chemistry syllabuses, it is hoped that it will be of interest to both chemists and biologists.

Questions taken from recent Advanced Level papers are included at the ends of chapters, and I am grateful to the University of London Schools Examinations Department, the Welsh Joint Education Committee, the Oxford and Cambridge Schools Examination Board, the Joint Matriculation Board, the Southern Universities Joint Board, the Associated Examining Board and the Northern Ireland Schools Examinations Council for permission to reproduce them. The material tested in these questions is not necessarily confined to the chapter at the end of which the question occurs because of the increasing tendency in recent years to set questions which test the candidates' ability to select material from different areas of the subject. The questions have therefore been placed at the ends of the chapters which seem most appropriate in each case. Another consequence of the increased prevalence of this type of question is that there are more questions which are appropriately placed at the ends of the later than of the earlier chapters.

Organic chemical nomenclature now presents some difficulties to the author of an Advanced level textbook. The dilemma is that on the one hand Examining

Boards have adopted systematic nomenclature, while on the other many substances are known almost universally amongst chemists by common names. In this book the recommendations on nomenclature by the Association for Science Education (which are largely similar to those of the International Union of Pure and Applied Chemistry) have been followed throughout, while for certain common substances common names are given as well. For a very few substances which are nearly always referred to by their common names, these names are used in the text after initial introduction of both common and systematic names. In this way it is hoped that the students will be introduced to systematic nomenclature and at the same time be enabled to read chemical literature in which the common names are used.

I should like to express my gratitude to Mr. Martyn Berry, who has read the manuscript, for many helpful suggestions and criticisms, and also to Mrs. Doris Storey for her invaluable assistance in preparing the manuscript.

1977 G.H.W.

Contents

Chapter 1

Basic Principles I : Covalent Bonds

Chemistry is concerned more with the making and breaking of bonds than with anything else. Since organic chemistry is the chemistry of carbon compounds, and since carbon characteristically takes part in covalent bonds, a study of organic chemistry must be based on an understanding of the nature of covalent bonds.

Atomic structure

The simple Bohr theory of atomic structure interprets the extra-nuclear electrons of atoms as occupying energy levels ('orbits', or 'shells', although none of these terms adequately describes the phenomenon), each of which can contain definite numbers of electrons: the first, two; the second, eight; the third, eighteen; and so on. This simple picture, which went some way towards explaining atomic phenomena, has been refined by the application of *wave mechanics*. Although it is still, like any theoretical approach, far from perfect, wave mechanics is capable of considerable refinement and has been notably successful in providing theoretical interpretations of the facts of chemical combination. A detailed account of wave mechanics would be out of place here, but it is necessary to summarize the two main conclusions which affect the carbon atom and its participation in covalent bonds.

The first conclusion (which is in fact a consequence of Heisenberg's 'Uncertainty Principle') is that it is physically impossible to define (or therefore to know) the precise position of an electron within an atom. All that it is possible to know is *the probability that the electron will be at a particular point at a particular time*, so that a three dimensional 'probability distribution map' can then be plotted. For the single electron in a hydrogen atom, for example, this probability distribution turns out to be spherically symmetrical, with the highest probability of occurrence of the electron about 0.053 nm from the nucleus and the probability falling off very rapidly with distance from the nucleus. This distance of maximum probability is about equal to the radius of the first Bohr orbit, and thus the wave mechanical picture can be seen as a development of this earlier one.

Variation of probability with distance from the nucleus ('radial distribution') is illustrated in Figure 1.1. Although the probability never quite reaches zero, because there must be a finite probability of finding the electron anywhere, it falls off very rapidly so that it is very small at distances much greater than the radius of the first Bohr orbit.

Since the electron is negatively charged this probability distribution is related to the density of negative charge, which, in a real sense, *is* the electron. Thus over a period of time, when the motion of the electron becomes 'blurred'

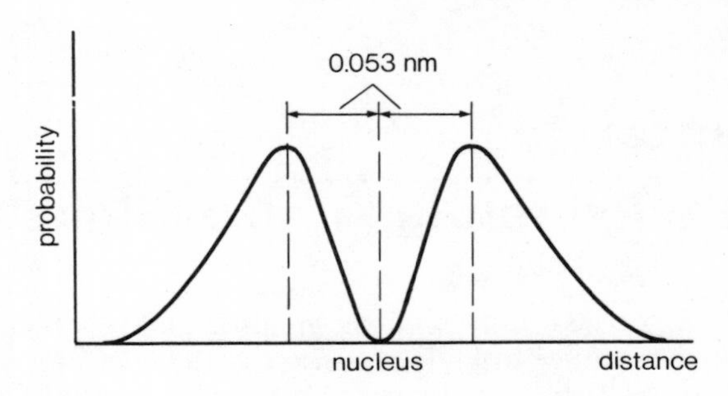

Figure 1.1

as in a time exposure, we can regard the electron as a 'cloud' of negative charge, the density of which at any point is related to the probability as plotted in Figure 1.1.

The function plotted in Figure 1.1 is in fact the charge density multiplied by $4\pi r^2$, where r is the distance from the nucleus. The charge density function itself is also spherically symmetrical, but its maximum is at the nucleus itself. If charge-density is represented by depth of shading, a two dimensional representation is as in Figure 1.2. This alternative mental picture can be useful, particularly in organic chemistry.

Thus the second main conclusion arising from wave-mechanical treatment is that the various electrons may be described by probability distribution (or charge density) maps. Moreover the various electrons in a given Bohr 'orbit' may be described by probability distribution or charge density maps *of different shapes*, and these are subject to definite rules.

These maps are known as *orbitals*, and we shall from now on refer to them as such. Each orbital has a definite potential energy associated with it, and orbitals are occupied by electrons in order of increasing potential energy. Thus only certain values of the energy of an electron is a given atom or molecule are permitted, and this energy can only increase or decrease in discrete amounts ('quanta'). It turns out that each orbital can accommodate up to two electrons, but no more than two. This is embodied in a principle

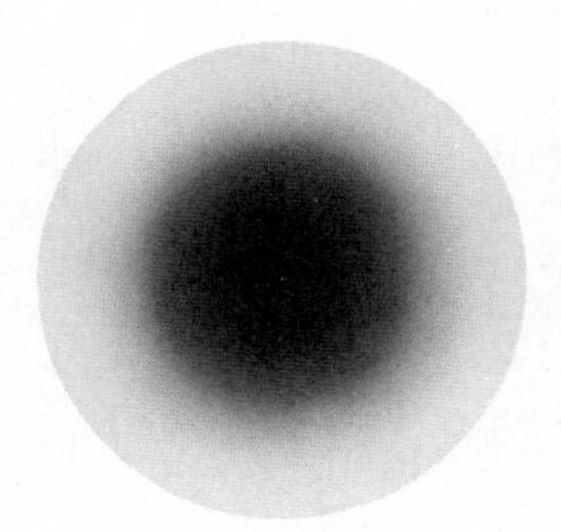

Figure 1.2

first stated by Pauli and known as the 'Exclusion Principle', which means in effect that no two electrons in an atom can be precisely the same. An electron behaves magnetically as a sphere with definite angular momentum as if it were spinning about an axis through its centre. Thus the only way in which electrons occupying an orbital, e.g. a spherically symmetrical one like that shown in Figure 1.1, can differ from one another is in respect of the directions of their spins. These spins can be opposite to one another, i.e. positive or negative, clockwise or anticlockwise, right or left, or any other *pair* of terms by which language describes opposites. There can, however, be no more than two which differ in this way. Two such electrons occupying any orbital are said to have their 'spins paired'.

The first Bohr orbit is labelled by the number 1 and this is known as its *main* or *principal quantum number*, n. The rule for $n = 1$ which emerges from the wave mechanical treatment is that it can contain only one orbital, the spherically symmetrical one just described. Such spherically symmetrical orbitals are known as s orbitals. This orbital can contain one or two electrons; with two it is full. In the hydrogen atom it contains one, and in the helium atom two with paired spins. The electronic configuration of hydrogen can be described as $1s^1$ (i.e. principal quantum number 1 and one s electron in this first Bohr orbit), and that of helium as $1s^2$ (i.e. principal quantum number 1 and two s electrons). This orbit is then full, and helium is a noble gas.

The third electron in the next element (lithium) must be placed in the next Bohr orbit ($n = 2$) and the orbital of lowest energy is again a spherically symmetrical s orbital. The electronic configuration of lithium is then denoted $1s^2\ 2s^1$, using the above conventional representation. The $2s$ orbital becomes filled in beryllium, which has the configuration $1s^2\ 2s^2$. The remaining six electrons to be accommodated in the second Bohr orbit require three additional orbitals. This is possible because, with $n = 2$, orbitals are permitted which have a directional quality, i.e. which are not spherically symmetrical. They are instead shaped like dumb-bells. In three-dimensional space there can be three distinct orbitals of this type, lying on the three (x, y, and z) axes mutually at right angles (*orthogonal*). They are known as the p_x, p_y, and p_z orbitals and are illustrated in Figure 1.3. Each p orbital has two maxima of

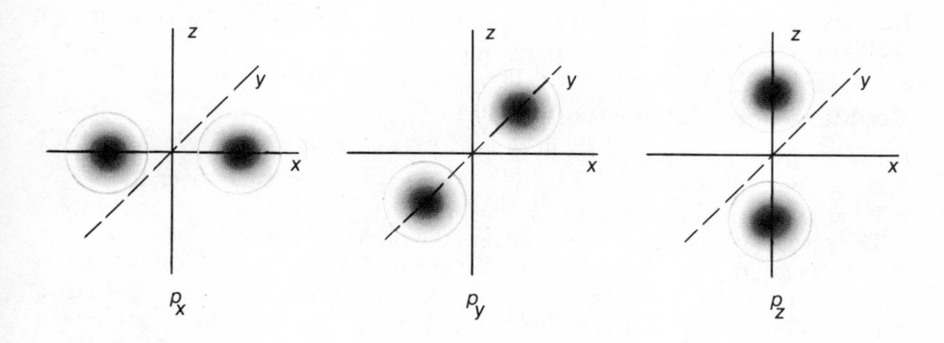

Figure 1.3

Table 1.1

	n = 1	n = 2				Configuration
	1s	2s	$2p_x$	$2p_y$	$2p_z$	
Hydrogen	↑					$1s^1$
Helium	↑↓					$1s^2$
Lithium	↑↓	↑				$1s^2\ 2s^1$
Beryllium	↑↓	↑↓				$1s^2\ 2s^2$
Boron	↑↓	↑↓	↑			$1s^2\ 2s^2\ 2p_x^1$
Carbon	↑↓	↑↓	↑	↑		$1s^2\ 2s^2 2p_x^1\ 2p_y^1$
Nitrogen	↑↓	↑↓	↑	↑	↑	$1s^2\ 2s^2\ 2p_x^1\ 2p_y^1\ 2p_z^1$
Oxygen	↑↓	↑↓	↑↓	↑	↑	$1s^2\ 2s^2\ 2p_x^2\ 2p_y^1\ 2p_z^1$
Fluorine	↑↓	↑↓	↑↓	↑↓	↑	$1s^2\ 2s^2\ 2p_x^2\ 2p_y^2\ 2p_z^1$
Neon	↑↓	↑↓	↑↓	↑↓	↑↓	$1s^2\ 2s^2\ 2p_x^2\ 2p_y^2\ 2p_z^2$

probability of occurrence of the electron and of charge density, lying on the x, y, or z axes respectively, with a plane of zero probability (a *nodal plane*) running through the nucleus itself. Each can accommodate two electrons, but it turns out that each accepts one electron before any accepts two. Thus boron has the configuration $1s^2\ 2s^2\ 2p_x^1$, carbon $1s^2\ 2s^2\ 2p_x^1\ 2p_y^1$, etc. The configurations of the first ten elements are set out in Table 1.1.

The next eight elements (sodium to argon) are placed analogously in the next Bohr orbit, using the 3*s* and 3*p* orbitals. For later elements, orbitals of another kind known as the *d* orbitals (of which there are five) become available, and later still, yet more complex *f* orbitals are needed. The letters *s*, *p*, *d*, and *f*, are the initial letters of certain spectral series, but the student may find it easier to remember if he thinks (perhaps with some justification) that this system of nomenclature was thought up by "some poor darn' fool"!

The *d* and *f* orbitals are not generally of much importance in elementary organic chemistry, but we shall be greatly concerned with *s* and *p* orbitals. Shorthand conventional representations such as those in Figure 1.4 are therefore useful; for *p* electrons both 'double circle' (a) and 'figure of eight' (b) diagrams are commonly used.

Covalent bond formation

If two hydrogen atoms are brought together the 1*s* orbitals will overlap and the electrons (of opposite spins) will be able to move freely over the larger volume of the combined orbital (Figure 1.5).

The resulting *molecular orbital* covers the molecule and now has symmetry about the interatomic axis. Such molecular orbitals are called *sigma orbitals* (σ *orbitals*), and the bond formed as a result of their occupation by an electron pair is called a σ bond. The bond energy (i.e. the gain in stability associated with bond formation) arises because the electrons are less confined in the molecular orbital than in the two atomic orbitals. It is a consequence of the

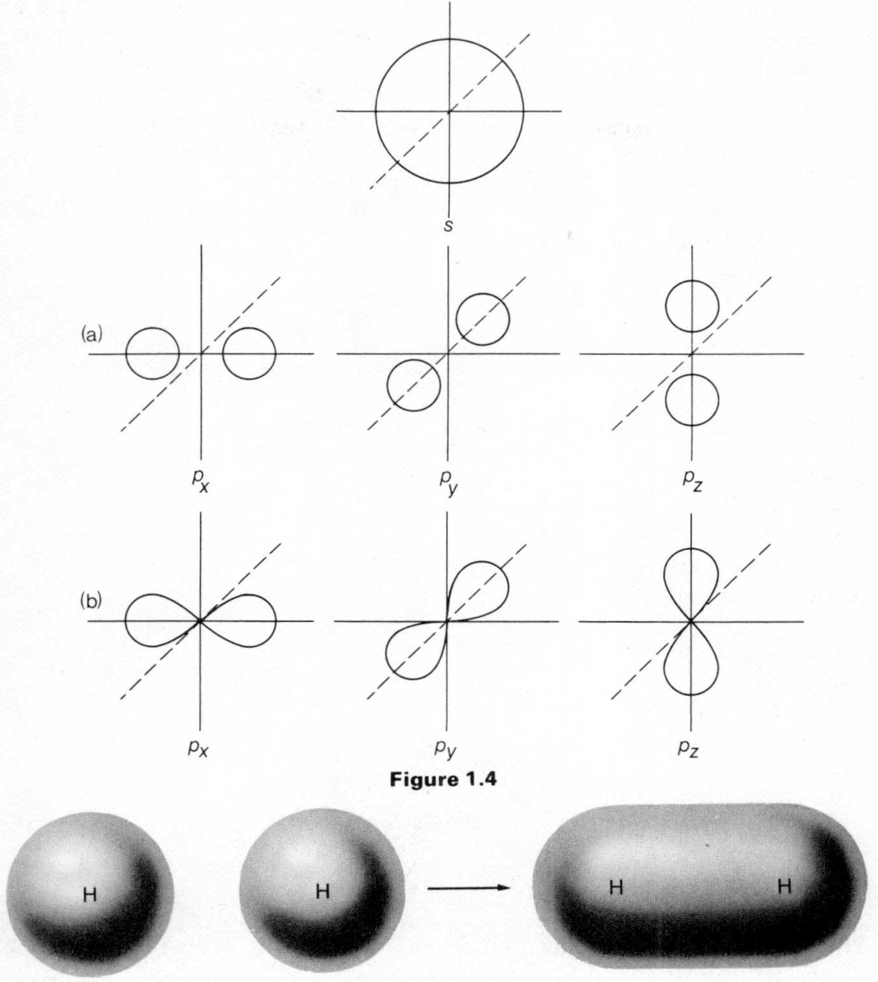

Figure 1.4

Figure 1.5

Uncertainty Principle that this *delocalization* of electrons leads to a reduction in energy. This is an important generalization and it will be met again later.

Carbon has only two unpaired electrons ($2p_x$ and $2p_y$) so it might be expected that carbon should be divalent, although we know it to be tetravalent. However, the most stable electronic arrangement in the carbon *atom* need not necessarily be the same as the most stable arrangement when bonds have been formed. Moreover, the orbitals can interact with one another, and this interaction means that they can be mixed together and then sorted out again in a different way as 'hybrid' orbitals. There are mathematical procedures for performing this treatment, and if they are applied to carbon atoms taking part in four covalent bonds the result is that the most stable arrangement is obtained as follows.

One of the two $2s$ electrons is promoted to the vacant $2p_z$ orbital, and the four electrons ($2s$, $2p_x$, $2p_y$, $2p_z$) *hybridized* as described above to give four hybrid orbitals of similar shape. These orbitals are called sp^3 hybrids and are shaped like lop-sided dumb-bells with one lobe much bigger than the other (Figure 1.6). Their orientation in space turns out to be such that the big lobes are directed towards the corners of a tetrahedron (Figure 1.7). This tetrahedron is geometrically regular if all four groups attached to the carbon atom are the same, but deviations from regularity may occur if they are not all the same.

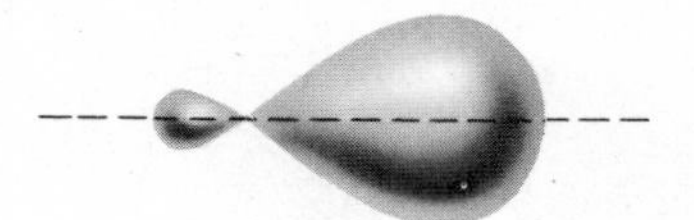

Figure 1.6 sp^3 hybrid orbital

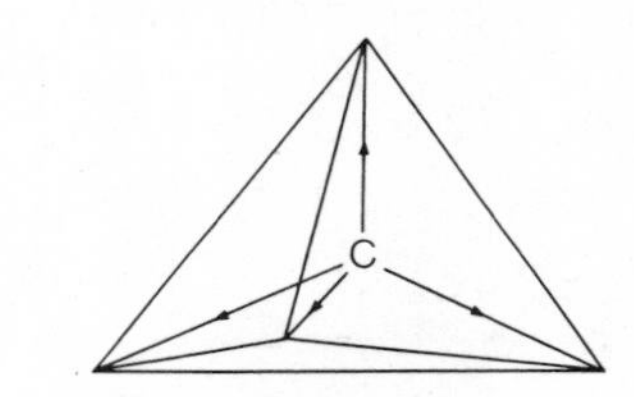

Figure 1.7

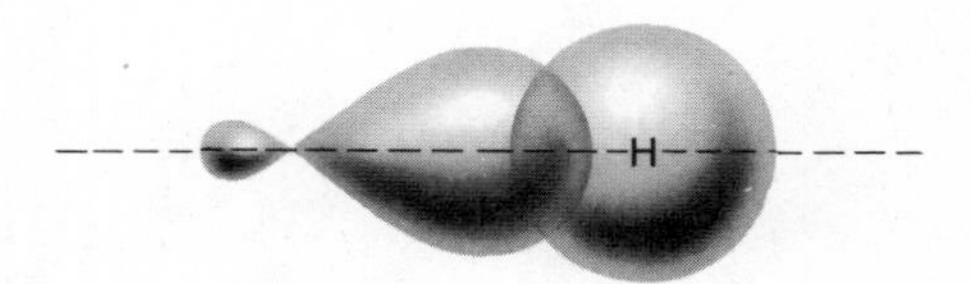

Figure 1.8

Bonds can be formed by overlap of the big lobes with other orbitals (e.g. a C–H bond as in Figure 1.8). The resulting bonds have symmetry about the interatomic axis and hence are σ bonds, and they are moreover directed towards the corners of a tetrahedron. This important conclusion about the shapes of molecules containing carbon atoms forming four bonds is experimentally justified (see Chapter 11, pages 116–39).

The same conclusion could also be reached from the simple and appealing idea that electron pairs repel one another. This tetrahedral arrangement is the one which puts the greatest distances between the four pairs in a molecule like methane, CH_4.

Multiple bonds

It is easy to see how bond formation as described above leads to compounds like CH_4, C_2H_6 (Figure 1.9) etc., but compounds like ethene (ethylene),

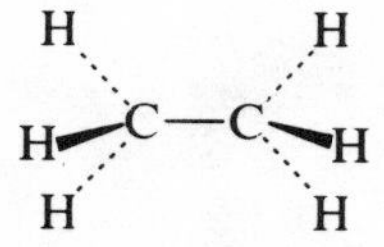

Figure 1.9

C_2H_4, in which the carbon atoms must be joined by a *double bond* involving the sharing of two pairs of electrons are also common.

The wave-mechanical picture of the double bond is that once again a $2s$ electron is promoted to the vacant $2p_z$ orbital, but that this time only the $2s$, $2p_x$, and $2p_y$ orbitals are hybridized, giving three sp^2 hybrid orbitals and leaving the $2p_z$ orbital unhybridized. The sp^2 orbitals are similar in shape to the sp^3 hybrids, but lie in a plane and are oriented towards the corners of an approximately equilateral triangle (i.e. trigonally disposed). The p_z orbital lies perpendicular to this plane. The C–H and C–C σ bonds in ethene are formed by overlap of the sp^2 orbitals with one another, and with the $1s$ orbitals of the hydrogen atoms respectively (Figure 1.10). This diagram shows the ethene molecule in plan, and conforms to our experimental knowledge of its geometry, i.e. planar with bond angles of 120°. The p_z orbitals lie above and below the plane of the molecule, which is their nodal plane. 'Sideways' overlap of these p_z orbitals gives a molecular orbital shaped rather like two fat sausages, one above and the other below the plane of the molecule, as shown in elevation in Figure 1.11.

Such a bond as this, formed by 'sideways' overlap of p orbitals, does not have symmetry about the interatomic axis like the σ bonds discussed above, and is known as a pi bond (π bond); the molecular orbital is a π orbital. Such π bonds have the important property that torsional twisting of the molecule about the interatomic axis cannot be accomplished without breaking the π bond. If one half of the molecule is rotated while the other is stationary, the lobes of the two p_z orbitals no longer overlap; energy must therefore be

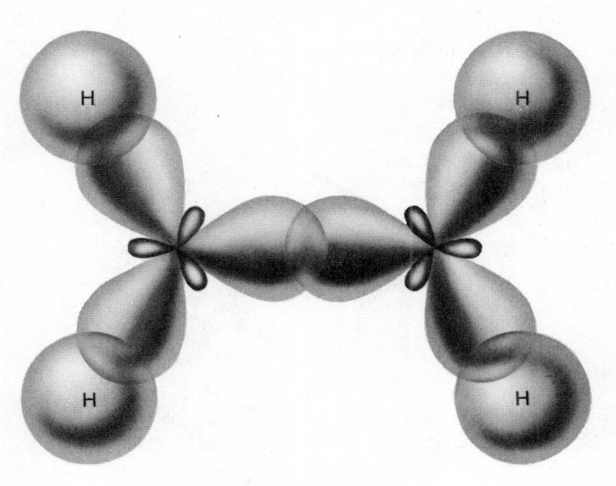

Figure 1.10

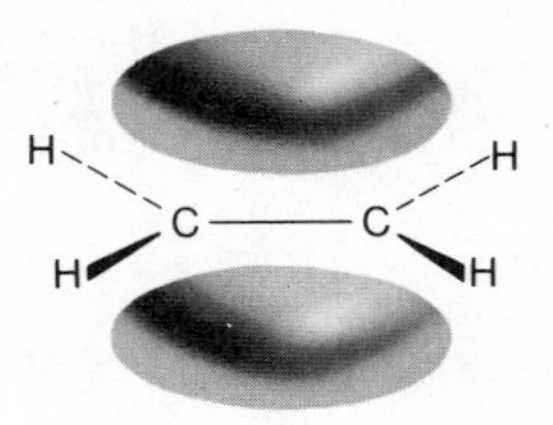

Figure 1.11

supplied in order to 'break' a π bond. This is why doubly bonded molecules are rigid, and whereas rotation about single bonds is normally free unless hindered for some other reason, it is not so about double bonds. Many properties and reactions of these 'unsaturated' compounds depend on the presence of the π bond.

The hydrocarbon ethyne (acetylene) C_2H_2 must contain a triple bond. This is formed similarly by hybridizing the $2s$ and $2p_x$ orbitals to give two *sp* hybrid orbitals to form the C–C and C–H σ bonds. The hybrids lie in a straight line and the geometry of the molecule is as in Figure 1.12.

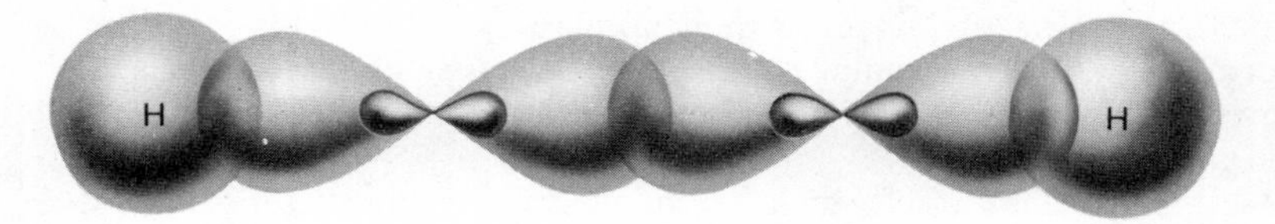

Figure 1.12

'Sideways' overlap now occurs of *two* pairs of *p* orbitals mutually at right angles, so that the molecule appears to be surrounded by two π orbitals, one in the *y*- and one in the *z*-direction, i.e. four 'sausages' of electron density. These coalesce to form a hollow cylinder of electron density, accommodating two pairs of electrons, surrounding the molecule. An end-on view is given in Figure 1.13.

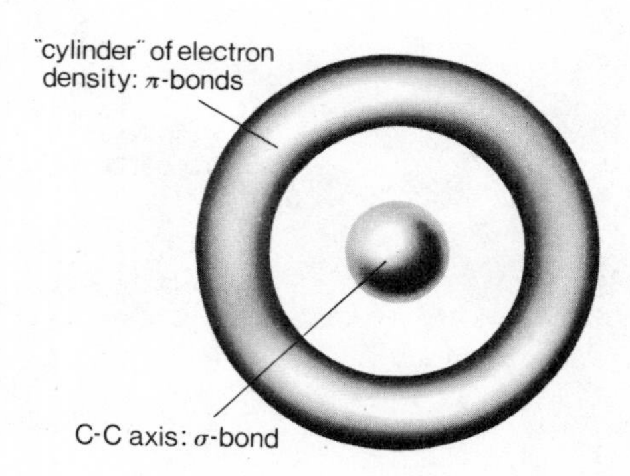

Figure 1.13

Molecular polarization

The fact that a pair of electrons is shared between two atoms to form a covalent bond does not necessarily mean that they are equally shared. If the bond is symmetrical, as in the hydrogen molecule, the sharing is, of course, equal; but if the electronegativities of the atoms forming the bond are considerably different, as they are for example in the HCl molecule, the bond electrons are more likely to be nearer the more electronegative atom (chlorine). The molecular orbital is distorted (Figure 1.14(a)) so that electron density is greater at the chlorine. The chlorine can be regarded as an electron-attracting atom, and the molecule becomes polarized in the sense shown in Figure 1.14(b), and is a dipole with a 'dipole moment' in this direction. [The symbols $\delta+$ and $\delta-$ ('delta-plus' and 'delta-minus') mean 'a very small amount of positive (or negative) charge'. The magnitudes of the charges represented by $\delta+$ and $\delta-$, if they were actual point charges centred at H and Cl respectively, would be much smaller than the electronic charge.]

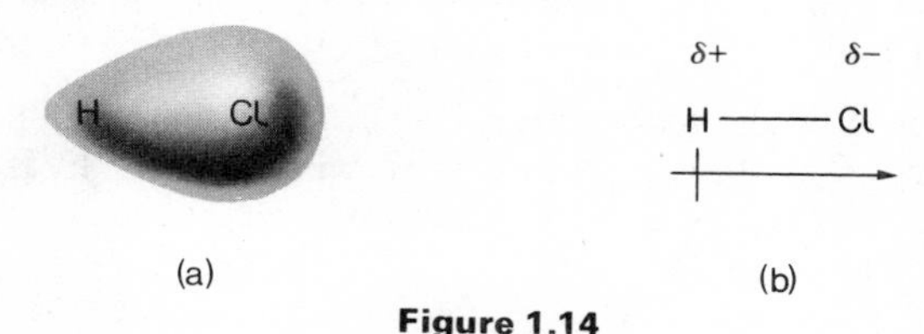

Figure 1.14

Chlorine can also attract electron density in this way when attached to a carbon chain, and the effect can be transmitted a little way along the chain by electrostatic induction, although it falls off very rapidly. The carbon atom (C1) adjacent to the chlorine becomes somewhat denuded of electron density by the effect of the chlorine, and itself then being poor in electron density, attracts electrons through the next bond from its neighbour (C2). Because of its transmission in this way, the effect is known as the *inductive effect* (symbolized as $-I$ for attraction and $+I$ for repulsion). Thus chlorine has an $-I$ effect. The $-I$ effects of atoms increase along a period of the periodic table with the electronegativities; thus, for $-I$ effects:

$$-NR_2 < -OR < -F.$$

The $-I$ effects of the halogens vary in the sense:

$$F > Cl > Br > I.$$

Carbon, and therefore alkyl groups, have a fairly small $+I$ effect of electron repulsion.

The inductive effect is one mechanism whereby atoms and groups can polarize molecules which contain them. Such molecular polarization, arising from unequal distribution of electron density in covalent bonds, has a profound influence on both physical and chemical properties.

The hydrogen bond

Many examples exist of the formation of weak links between two electronegative atoms like fluorine, oxygen, or nitrogen (which may be in the same or in different molecules) through a hydrogen atom bound to one of them. It is by means of such *hydrogen bonds* that water and ammonia, for example, become extensively associated in the liquid phase.

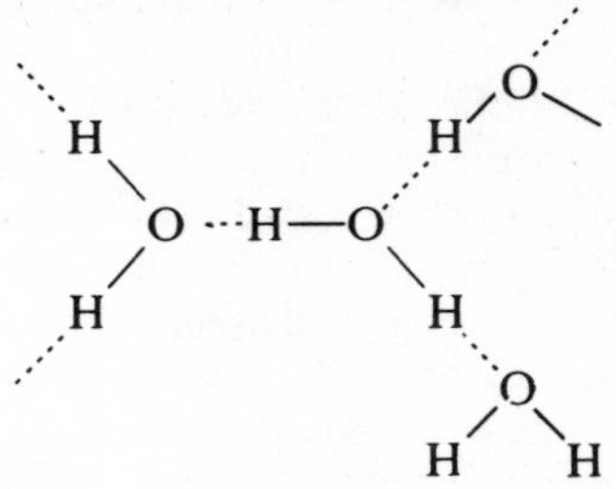

(Hydrogen bonds are usually represented by broken lines.) The energy associated with them varies a good deal according to the participating atoms and their molecular environments. For –O---H–O bonds it is usually of the order of 25–30 kJ mol^{-1}. The source of this energy decrease on formation of a hydrogen bond is thought to be largely electrostatic. The $\overset{\delta-}{O}$—$\overset{\delta+}{H}$ bond, for example, is polarized as shown, and the somewhat positive hydrogen atom is obviously attracted electrostatically to another electronegative atom in its vicinity.

Question

1. 'Chemical bonding, irrespective of whether it is ionic or covalent, is basically the result of electrostatic interactions between electrons and nuclei. The essential difference between the two types of bonding lies in the distribution of electrons.'
Explain and discuss this statement.

(London, Special Paper, 1973)

Chapter 2

Basic Principles II : Kinetics and Mechanisms

Why study reaction mechanisms?

Organic compounds undergo many reactions, some simple, some highly complex and unexpected. A knowledge of the subject involves a knowledge of these reactions, and unless that knowledge is based on an understanding of how (and often, therefore, why) these reactions take place, learning organic chemistry degenerates into learning a mass of unrelated facts. This understanding comes from a study of reaction mechanisms, which poses questions such as: Does the reaction proceed smoothly from reactants to products, or are there several stages involved? If there are several stages, which are fast and which are slow? What bonds are made and broken, and in what order? Do the electron pairs stay together, or are they split up so that unpaired single electrons are involved? How do the electrons move about the molecules during the reaction?

The understanding of a reaction mechanism also enables us to use the reaction to much better effect. For example, it may enable us to predict the effects of *constitutional* changes (e.g. changes in substituent groups present or changes in the nature of the reagent) or *environmental* changes (e.g. changes in the solvent, or the presence of added substances). Hence we may be able to choose the best conditions under which to conduct the reaction, to design variations on it, or to develop new reactions analogous to it. Such developments usually depend on a precise knowledge of that stage in a complex, multi-stage reaction which determines the rate of the whole process. This so-called *rate-determining stage* is the slowest of the several stages. The rate of a reaction is expressed as moles per cubic decimetre (moles per litre) of products formed (usually the same as 'reactants consumed') per second and its units are therefore mol dm^{-3} sec^{-1} (mol l^{-1} sec^{-1}). The rate at which products are formed is obviously determined by the rate of the slowest stage. A simple analogy would be a chain of people passing buckets of water from one to the other for the end person to pour on a fire. The rate at which buckets reach the fire is determined by the rate at which the slowest person in the chain passes them, because he holds up all those after him.

The investigation of reaction mechanisms

Many methods are used to study reaction mechanisms, and as many as possible will be illustrated in this book, but the main weapon in our armoury is reaction kinetics, i.e. a study of the rates of reactions under various conditions. We can find out which reactants affect the rate and how this is affected by various constitutional and environmental factors, and hence make deductions about the mechanism of the reaction. Clearly there is not time in a course such as this (or in any other course for that matter) to study in detail the experimental evidence and the reasoning which has led to our knowledge of every reaction mechanism we meet. Indeed in most cases it will be necessary to take this on trust, and to accept that a reaction proceeds in such and such a manner. Very often the mechanism will appear clearly rational and attractive intellectually. The student must be on his guard, however, that he does not get the idea from such attractiveness that one only has to think about a reaction for a few minutes to be able to work out its mechanism. It cannot be emphasized too strongly that the many mechanisms which we now take for granted were established only by painstaking kinetic and other investigations.

While many mechanisms in this book will be given without justification, to illustrate the point two classic examples have been selected for more detailed treatment in respect of the investigative work which has led to their establishment (see pages 61–6 and 79–82).

Kinetic concepts

It will be useful at this point to define and explain a few of the fundamental concepts of reaction kinetics.

Rate of reaction

We have already defined this as rate of change of concentration, and if we are considering it as the rate of consumption of a reactant A, the rate at any given instant (the instantaneous rate) is $-\mathrm{d}[\mathrm{A}]/\mathrm{d}t$. The minus sign is there because we are considering a *decrease* in concentration, and $\mathrm{d}[\mathrm{A}]/\mathrm{d}t$ is itself negative. A negative sign is therefore needed to give a positive rate. If we were considering appearance of a product, we would not need the negative sign.

Order of reaction

Reaction rates are proportional to 'activities' (concentrations, to a first approximation) of reactants. We have to find out experimentally which concentrations affect the rate, and how. The order of a reaction is an expression of the results of these experiments, and is the number of concentration terms to which the rate is found *experimentally* to be proportional. Thus if one concentration term is involved the reaction is first order, if two are involved (either as different species, or as the square of the concentration of a single species) it is second order, etc.

First order

$$-\frac{\mathrm{d}[\mathrm{A}]}{\mathrm{d}t} \propto [\mathrm{A}] \qquad (1)$$

Second order

$$-\frac{\mathrm{d}[\mathrm{A}]}{\mathrm{d}t} \propto [\mathrm{A}]\,[\mathrm{B}] \qquad (2)$$

$$-\frac{d[A]}{dt} \propto [A]^2 \tag{3}$$

In these equations, and throughout the book, square brackets signify concentrations; thus [A] means 'the molar concentration of the species A'.

If a reaction of the type of (2) were to be conducted under conditions such that B was present in very large, say 100-fold, excess over A, by the time the entire reaction had proceeded, i.e. *all* of A had been consumed, the concentration of B would have changed by only one per cent and the influence of B at the end of the reaction would be 99/100 of its influence at the beginning. Such a small change would not be observed experimentally in the presence of a much larger change in the influence of A, so the experimental result would be that, *under these conditions*, B does not influence the rate and the reaction becomes first-order, obeying equation (1). This trick, which can be used to simplify the kinetics by reducing the order, is known as Ostwald's Isolation Method. It has been given here to emphasize the fact that order of reaction is an *experimental* result; it is *not* characteristic of the reaction, but can vary according to the conditions under which the reaction is carried out, although the mechanism need not vary.

Rate constant

This is simply the proportionality constant in (1), (2), or (3).

First order

$$-\frac{d[A]}{dt} = k_1 [A] \tag{4}$$

Second order

$$-\frac{d[A]}{dt} = k_2 [A] [B], \text{ etc.} \tag{5}$$

Here k_1 is called a first order constant, and k_2 a second order constant. By examining the dimensions of the quantities in (4) and (5) it can be seen that k_1 has the dimensions sec^{-1}, and k_2, $dm^3\ mol^{-1}\ sec^{-1}$ (litres $mol^{-1}\ sec^{-1}$).

The rate constant is, of course, a measure of the *reactivity* of the reagents (i.e. how reactive they are), but it cannot by itself tell us how fast a reaction will proceed, because the rate also depends on the concentrations of the reagents. If an intermediate stage is rate-determining, one of these reagents may be an intermediate (possibly an ion or a free radical) which may be present in an unknown concentration. Thus although such an intermediate may be extremely reactive (i.e. the rate constants may be very high), unless it is available in sufficient quantity (concentration) to do its job, the reaction will not proceed rapidly.

Measurement of rate constants

Reactions are usually followed by measuring the concentration (or some physical property which is proportional to the concentration) of a reactant at given intervals of time. For a first-order reaction, if [A] is plotted against time, an exponential variation is observed (Figure 2.1).

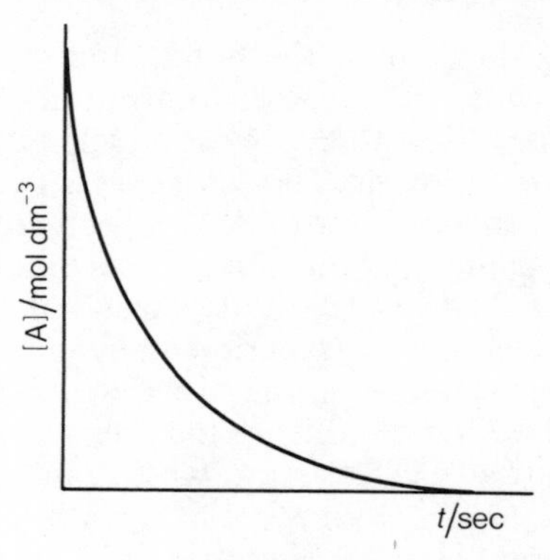

Figure 2.1

Instantaneous rates could of course be obtained as the slopes of tangents to this curve and k_1 obtained from (4), but this is highly inaccurate, and it is necessary to find a linear function to plot. Equation (4) must therefore be integrated.

Rearrangement of (4) gives

$$-\frac{d[A]}{[A]} = k_1 \, dt$$

and integration gives

$$-\ln [A] = k_1 t + C$$

At $t = 0$, $-\ln [A]_0 = C$, where $[A]_0$ is the initial concentration of A, so

$$\ln [A] = k_1 t + \ln [A]_0$$

or,

$$\log [A] = \frac{k_1 t}{2.303} + \log [A]_0$$

Thus if log [A] is plotted against t, a straight line is obtained (Figure 2.2).

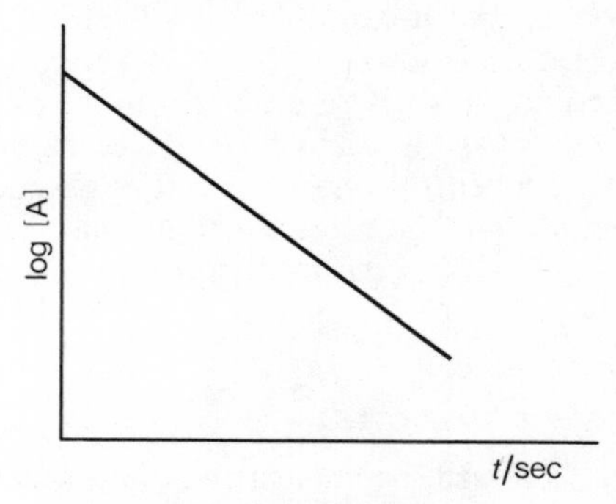

Figure 2.2

The intercept is log $[A]_0$ and the slope is $-k_1/2.303$, from which k_1 can be obtained.

Integration of second order rate equations is more complicated, but straight lines can again be obtained from appropriate plots and the rate constants measured similarly. However, it is usually best if possible to use Ostwald's method to reduce the reaction order to unity and to measure first order constants.

In a few cases, the operation of the isolation technique can result in a reaction order being reduced to zero. The rate is then independent of any concentrations, and in obedience to equation (7) the rate is constant. The rate constant is now the same as the rate itself (k_0)

$$-\frac{d[A]}{dt} = k_0 \tag{7}$$

and the graph of [A] against t is a straight line of slope k_0 with a sharp cut-off when all of A has been consumed (Figure 2.3).

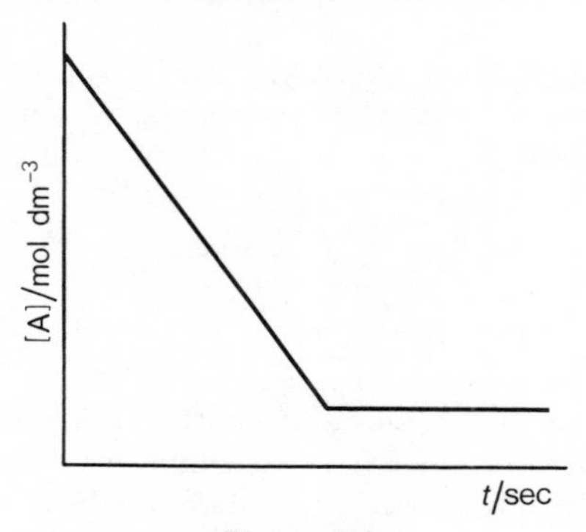

Figure 2.3

The achievement of zero order kinetics can be extremely useful in deducing a reaction mechanism (cf. Chapter 7, pages 58–72).

Molecularity

The molecularity of a reaction is defined as the number of molecules taking part in (or better, undergoing covalency changes in) the rate-determining stage. Thus a one-stage decomposition reaction of type (8) is unimolecular,

$$A \longrightarrow B + C \tag{8}$$

and a one-stage process of type (9) is bimolecular.

$$A + B \longrightarrow \text{products} \tag{9}$$

However, if a reaction conventionally expressed by (9) actually proceeds in stages (10) and (11), it is unimolecular, because only A participates in the rate determining stage (10). Such a reaction is normally first order. The bimolecular process (9) can be seen to be second order if the influence of the concentrations

$$A \xrightarrow{\text{slow}} C \quad (10)$$

$$B + C \xrightarrow{\text{fast}} \text{products} \quad (11)$$

of both A and B are observable. However, if B is in great excess it will be first order, although it is still bimolecular. This can arise if, for example, B is a main constituent of the solvent (e.g. water in an aqueous solvent). In such a reaction (which is called *solvolysis*) this is unavoidable, and often conditions cannot be found in which variation in [B] is significant and second-order kinetics cannot be observed.

Thus there is no simple and universal relationship between order and molecularity, although information concerning reaction order, if correctly interpreted, can lead to a knowledge of molecularity. Solvolysis is the most difficult case, because the kinetics are first order whether the reaction is bimolecular (9) or unimolecular (10 and 11) and kinetics alone are not sufficient to diagnose the molecularity. Other information, of different kinds, is necessary as well.

The difference between order and molecularity is essentially that order is a piece of *experimental* information and is dependent on the way in which the reaction is carried out. Molecularity is more fundamental. It is directly related to the mechanism of the reaction, a *theoretical* concept, and unless the mechanism changes is independent of the way in which the reaction is carried out.

The Transition State

If we make a hypothetical plot of the energy of a molecule or molecules undergoing reaction against the extent to which reaction has proceeded (the so-called 'reaction co-ordinate') a graph such as Figure 2.4 will be obtained for an exothermic reaction and Figure 2.5 for an endothermic reaction. The enthalpy of reaction, ΔH, is the difference between the enthalpies of the initial and final states (reactants and products). The state corresponding to the energy maximum is the so-called 'transition state' and the difference, E, between the energy maximum and the reactants is the 'activation energy'. This is the minimum energy which must be supplied to a molecule to enable it

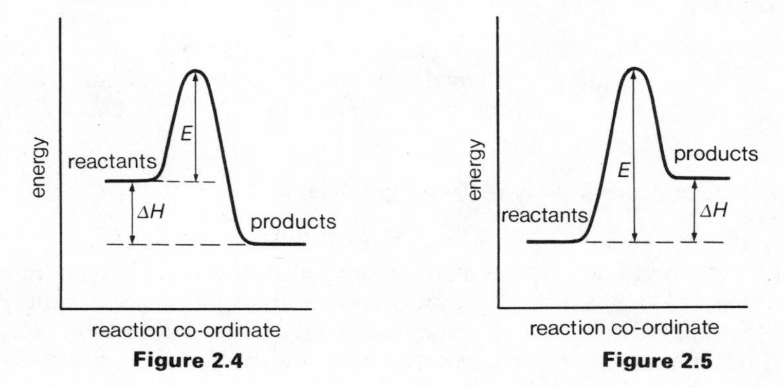

Figure 2.4 **Figure 2.5**

to react (to climb the hill between the two valleys). The structure of the transition state depends on the reaction, and a knowledge of it is an essential part of understanding the mechanism. It always contains partially formed and partially broken bonds and is the most unstable (i.e. most energetic) state through which the reactants pass on their way to the products. The activation energy is a most important parameter, since it influences the rate of the reaction in an inverse sense: the higher the activation energy, the lower the rate, and vice versa. A multi-stage reaction has several transition states. A system like (10) and (11) in which the first stage is rate-determining, for example, is represented by Figure 2.6. The well in the potential energy curve represents the intermediate, and the maxima the transition states. E_1 is the activation energy of the first stage and E_2 that of the faster second stage. E_1 is also the experimentally determined activation energy because the first stage is rate-determining.

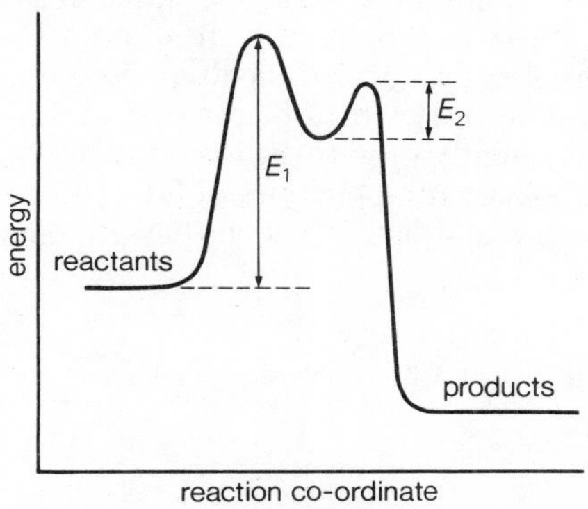

Figure 2.6

Measurement of activation parameters : the Arrhenius equation

The activation energy is measured by observing the variation of the rate constant with temperature. Rate constant and temperature are related by the Arrhenius equation (12),

$$k = A\,e^{-E/RT} \qquad (12)$$

where E is the activation energy, R the gas constant, and T the absolute temperature. A is a proportionality constant sometimes known as the non-exponential factor. It is related to the entropy of activation (i.e. the difference in entropy between the initial and transition states) because the greater the increase (or the smaller the decrease) in entropy involved in forming the transition state, the faster the reaction will proceed. This can sometimes be an important factor, for example when the formation of a transition state involves fitting in a bulky reagent to an already crowded molecule. The transition state then needs to be highly organized because there will be a strict limitation on the ways in which all the atoms and groups can be arranged in space, and there will have to be a considerable decrease in entropy (i.e. a decrease in disorder, or an increase in organization) in order to form the transition state. This effect will make it difficult to form the transition state and the reaction will be correspondingly slow.

Such cases apart, however, we are usually much more concerned with the activation energy as a factor influencing rate, and most of our discussions of activation parameters are concerned with E rather than A.

If logarithms are taken of both sides of (12) we obtain

$$\ln k = \ln A - \frac{E}{RT}$$

or

$$\log k = \log A - \frac{E}{2.303\ RT} \qquad (13)$$

Thus if $\log k$ is plotted against $1/T$ a straight line is obtained whose intercept on the ordinate is $\log A$ and whose slope is $-E/2.303\ R$. Thus A and E can be obtained. The rate of reaction should be measured at as many temperatures as possible—at least three and preferably four or five, at 5 or 10 degree intervals. This can involve practical difficulties because, depending on the value of E, the rate usually varies by a factor of 2 to 3 per 10 °C so that the reaction may be inconveniently slow at the lowest temperature, or unmanageably fast at the highest. Activation energies of 80–120 kJ mol^{-1} are common although there are many reactions for which they are outside this range.

Question

1. In the presence of hydrochloric acid, *N*-chloroacetanilide (A) is changed to its isomer, 4-chloroacetanilide (B):

C_6H_5—N(Cl)—CO . CH_3 ⟶ Cl—C_6H_4—NH . CO . CH_3.

A *B*

The progress of the reaction can be followed because A liberates iodine from potassium iodide solution whereas B does not react. The iodine can be estimated by titration with standard sodium thiosulphate solution.

The table shows the volume, x, of a sodium thiosulphate solution needed at various times in the course of a particular experiment, in each case for a fixed volume of reaction mixture. (x measures the amount of A left at each titration.)

t (min)	0	15	30	45	60	75
x (cm^3)	24·5	18·1	13·3	9·7	7·1	5·2

(*a*) Using suitable scales, plot a graph of x against t.

(*b*) From your graph, read off the time taken
(i) for half of the original A,
(ii) for three quarters of the original A, to have been transformed into B.
What does a comparison of these two figures tell us about the order of the reaction?

(*c*) Describe in outline how you would investigate the effect of changing acid concentration on the rate of the reaction.

(*d*) Discuss briefly the effect that an increase in temperature would have on the rate of the reaction.

(*e*) Does the fact that a reaction is of first order necessarily mean that it is unimolecular? Explain.

(London, 1973)

Chapter 3

Frameworks and Functional Groups

Carbon atoms can be joined together in different ways to form frameworks or skeletons to which other atoms or groups can be attached. If all these are hydrogen atoms, the compounds are *hydrocarbons*, substances composed of carbon and hydrogen only. Other groups attached in place of hydrogen are known as *functional groups* and are characteristic of the various families of organic compounds. Thus, for example, all *alcohols* contain the hydroxyl group, –OH. If one hydrogen in a hydrocarbon is replaced by a functional group, the rest of the molecule (the framework to which the functional group is attached) can be designated by the suffix *-yl*. Thus the hydrocarbon CH_4 is methane and the group CH_3– is the methyl group. The functional groups have characteristic reactions, but these may often be influenced or modified by the frameworks to which they are attached. Likewise reactions of the frameworks themselves may depend on the functional groups attached. This interaction between functions and frameworks is a recurring theme in organic chemistry, and many examples of it will be found in this book.

Saturated carbon frameworks

Frameworks containing only sp^3 carbon, i.e. with no multiple bonds, are called *saturated*. It will be recalled that the four bonds formed by sp^3 carbon are oriented in space towards the corners of a regular tetrahedron, so to give a true picture of the geometry of the methane molecule, CH_4, should be drawn as in (*a*) below. The hydrogen atoms at the thick ends of the bonds drawn as

```
     H                    H
     |                    |
     C               H—C—H
   ◢ ⋮ ◣                  |
  H  H   H                H

    (a)                  (b)
```

heavy wedges lie above the plane of the paper, while the hydrogen atom at the end of the bond drawn as a broken line lies below the plane of the paper. The central carbon atom and the fourth hydrogen atom (that at the end of the bond drawn as a plain line) lie in the plane of the paper.

It is often necessary to use 'perspective' formulae of this kind, but unless the spatial relationships need to be illustrated it is more usual to use a 'shorthand reduction' of the formula to two dimensions, as in (*b*). It should, however, always be remembered that such formulae are only a conventional representation of formulae like (*a*) and that the angles between the bonds are not 90°, but about 109.5° (the 'tetrahedral angle'). These C–H bonds are about

0.112 nm long and the dissociation energy D (CH_3–H), i.e. the enthalpy of the reaction is 426.8 kJ mol^{-1}.

$$CH_4 \longrightarrow CH_3\cdot + \cdot H$$

If two carbon atoms are joined together the compound C_2H_6 (ethane) is obtained, and with three carbons C_3H_8 (propane).

```
    H   H              H   H   H
    |   |              |   |   |
H—C—C—H          H—C—C—C—H
    |   |              |   |   |
    H   H              H   H   H

   ethane              propane
```

It is often not necessary to give the structures in as much detail as this; the main features of the carbon framework can be indicated thus:

CH_3CH_3 (ethane) $CH_3CH_2CH_3$ (propane)

When we come to a four-carbon system we find something new: there are two different ways of joining four carbon atoms together. There are therefore two possible structural arrangements which correspond to the molecular formulae C_4H_{10}.

```
    H   H   H   H
    |   |   |   |
H—C—C—C—C—H      or      CH3CH2CH2CH3
    |   |   |   |
    H   H   H   H
```

and

```
H      H H      H              H3C
 \    /   \    /                  \
   C        C          or          CH—CH3
 /    \   /    \                  /
H      CH       H              H3C
       |
     H—C—H
       |
       H
```

These two different structures correspond to two different compounds, which have quite distinct physical and chemical properties. Compounds related to one another in this way (i.e. which have the same molecular formula but different structures) are known as *isomers* and are said to be *isomeric* with one another. There are various kinds of isomerism which are sometimes given special names, and which will be met from time to time.

The isomerism here arises from the possibility of chain branching. In the next member of the hydrocarbon series, C_5H_{12}, there are three isomers arising from chain-branching. The number of isomers increases rapidly as this series is ascended.

$$CH_3CH_2CH_2CH_2CH_3$$

$$(H_3C)_2CHCH_2CH_3 \quad \text{or} \quad (CH_3)_2CHCH_2CH_3$$

$$H_3C{-}C(CH_3)_2{-}CH_3 \quad \text{or} \quad (CH_3)_4C$$

We have begun to refer to a series of these hydrocarbons, and such series are known as *homologous* series. Their members are *homologues* of one another. It will have been noticed that the members of this series can be expressed by a general formula $C_n H_{2n+2}$, and that the increment in going from one member to the next is CH_2. Members of a homologous series have roughly similar *chemical* properties, and generally show a *gradation* of *physical* properties. Some physical properties of the simpler straight-chain members of this series are given in Table 3.1.

Table 3.1

	Melting point °C	*Boiling point* °C	*Relative density (as liquids)*
Methane, CH_4	−183	−162	0.424
Ethane, C_2H_6	−172	−89	0.546
Propane, C_3H_8	−187	−42	0.582
Butane, C_4H_{10}	−135	−0.5	0.579
Pentane, C_5H_{12}	−130	36.1	0.626
Hexane, C_6H_{14}	−94	68.7	0.669

The melting points are a little irregular at first, but later rise regularly at a decreasing rate. Boiling points rise regularly, again at a decreasing rate. It may be useful to remember that the straight-chain member of relative molecular mass 100 (heptane, C_7H_{16}) boils at 100 °C. The relative densities rise at a decreasing rate to a maximum of about 0.8, so all these compounds are less dense than water. The straight chain members C_1 to C_4 are gases at room temperature, C_5 to C_{17} are liquids, and from C_{18} onwards they are solids.

Chain branching also affects physical properties. For example, the higher the degree of branching, the more compact the molecule becomes and this usually results in a lowering of the boiling point. This is illustrated for the hydrocarbons C_5H_{12} in Table 3.2.

Hydrocarbons belonging to this series of general formula C_nH_{2n+2} are

Table 3.2

	Boiling point °C
$CH_3CH_2CH_2CH_2CH_3$	36.1
$(CH_3)_2CHCH_2CH_3$	28
$(CH_3)_4C$	9.5

known as *alkanes*, or sometimes *paraffins*. The monovalent framework groups C_nH_{2n+1} are called *alkyl groups* (e.g. CH_3, methyl; C_2H_5, ethyl, etc.).

It is possible to arrange the carbon atoms in rings with various numbers of members, e.g. C_6H_{12}, cyclohexane.

CH_2—CH_2, H_2C, CH_2, CH_2—CH_2 H_2C, CH_2, CH_2, CH_2, H_2C, CH_2

cyclohexane

This six-membered ring is not flat, because puckering as shown allows the bond angles to approach the tetrahedral angle more closely. The molecule is therefore less *strained*, and consequently more stable.

This is the so-called 'chair' form of cyclohexane; there is another form, called the 'boat' form, which is somewhat less stable.

H_2C CH_2
CH_2—CH_2
CH_2—CH_2

cyclohexane, 'boat' form

Both forms are readily interconvertible, with only a low activation energy, so 'boat' cyclohexane cannot be isolated at ordinary temperatures. Under normal conditions, cyclohexane and its derivatives exist almost entirely in the chair form.

Alternative forms like this which can be inter-converted by twisting about bonds without breaking those bonds are called alternative *conformations* of the molecule. Sometimes strain is unavoidable in forming a ring, as in C_3H_6, cyclopropane, and the compound then suffers a corresponding instability. The general formula of these *cycloalkanes* is C_nH_{2n}.

Unsaturated carbon frameworks

The series of hydrocarbons containing a double bond is the series of *alkenes*, or *olefins*, the first member of which is ethene (ethylene), C_2H_4. The general formula is C_nH_{2n} so the alkenes are isomeric with the cycloalkanes.

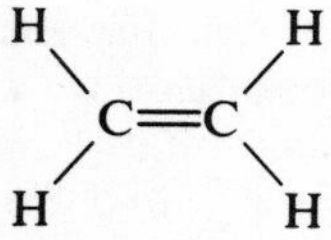

```
H       H
 \     /
  C=C
 /     \
H       H
```

or $CH_2{=}CH_2$ or $CH_2{:}CH_2$

ethene (ethylene), C_2H_4

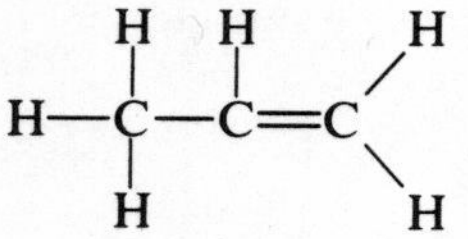

```
   H  H       H
   |  |      /
H—C—C=C
   |         \
   H          H
```

or $CH_3{-}CH{=}CH_2$ or $CH_3CH{:}CH_2$

propene (propylene), C_3H_8

Isomerism arises in the butenes, C_4H_8, because of alternative positions for the double bond, as well as because of chain branching.

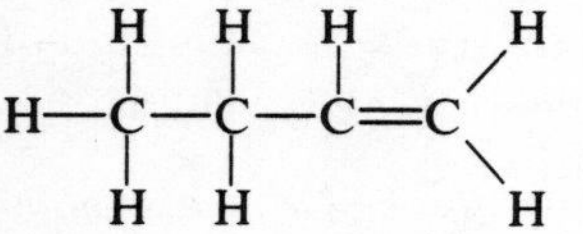

```
   H  H  H       H
   |  |  |      /
H—C—C—C=C
   |  |         \
   H  H          H
```

or $CH_3{-}CH_2{-}CH{=}CH_2$ or $CH_3CH_2CH{:}CH_2$

but-1-ene

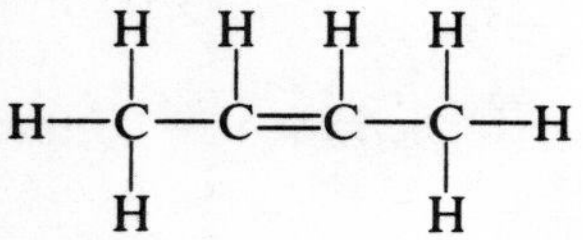

```
   H  H  H  H
   |  |  |  |
H—C—C=C—C—H
   |        |
   H        H
```

or $CH_3{-}CH{=}CH{-}CH_3$ or $CH_3CH{:}CHCH_2$

but-2-ene

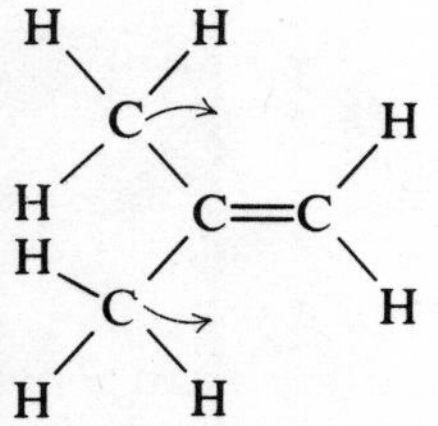

```
H   H
 \ /
  C          H
 / \        /
H    C=C
H   /       \
 \ C         H
  / \
 H   H
```

or

```
H3C
    \
     CH=CH2
    /
H3C
```

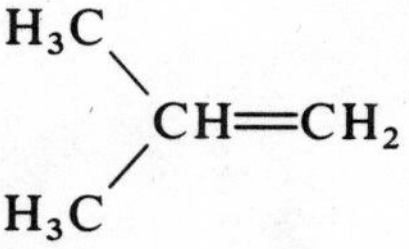

or $(CH_3)_2CH{:}CH_2$

2-methylpropene*

The geometry of sp^2 hybridized carbon was dealt with in Chapter 1. It will be remembered that the bond angle is 120°, the three bonds being coplanar. It follows that ethene is a flat molecule and the structure as drawn above is a

*The naming of these and other organic compounds is subject to rules, which are dealt with when the various series are discussed in detail.

correct representation. The homologues are not planar because they contain sp^3 as well as sp^2 carbon atoms, although the double bond and the four other bonds formed by the doubly bound carbon atoms are coplanar.

It will be remembered that because of the overlap of the p_z orbitals (Chapter 1, pages 7–8), carbon atoms forming a double bond (with their attached groups) cannot be rotated with respect to one another about the double bond. Thus structure I cannot be converted into structure II without supplying a good deal of energy.

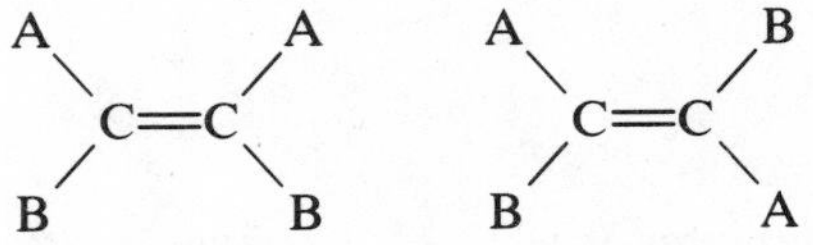

I(*cis-*) II(*trans-*)

These two structures are therefore isomers; they are physically and chemically different, and are known as the *cis-* and *trans-* forms respectively. This kind of isomerism is called *geometrical isomerism*, and will be met later (Chapter 11, page 137).

It is possible for there to be more than one double bond in the molecule. Hydrocarbons of this kind are called *polyenes*, e.g. $CH_2{:}CHCH{:}CH_2$, buta-1,3-diene.

The double bond itself can be regarded as a functional group with its own characteristic reactions, e.g. *addition* (cf. Chapter 4), as in the reaction of ethene with bromine

$$CH_2{:}CH_2 + Br_2 \longrightarrow CH_2Br.CH_2Br$$

Double bonds can occur in cyclic compounds, and when they do so they preserve their geometry as far as possible. The shape of the cyclohexene, C_6H_{10}, molecule is therefore

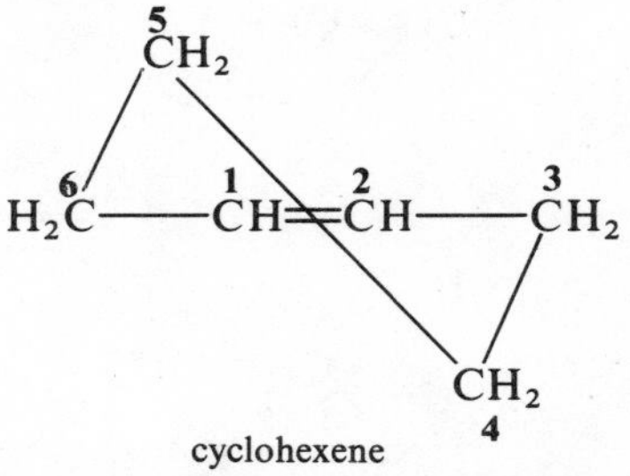

cyclohexene

Atoms 1, 2, 3, and 6 lie in one plane, while atoms 4 and 5 are below and above that plane respectively. (In this diagram we are looking at the molecule with the eye in the 1, 2, 3, 6 plane.) More frequently cyclohexene is conventionally represented as

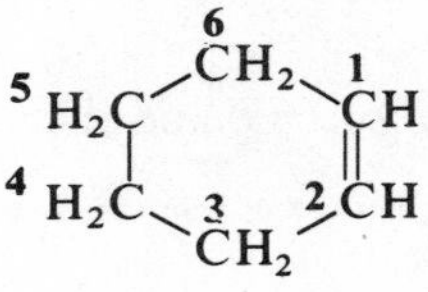

The simplest compound containing a *triple* bond is ethyne (which for many years was normally called acetylene), C_2H_2, and it is the first member of the series of alkynes, which have the general formula C_nH_{2n-2}. The next member is propyne (methylacetylene), C_3H_4,

$$H{-}C{\equiv}C{-}H \quad \text{or} \quad CH{\equiv}CH \quad \text{or} \quad CH{:}CH$$

ethyne (acetylene), C_2H_2

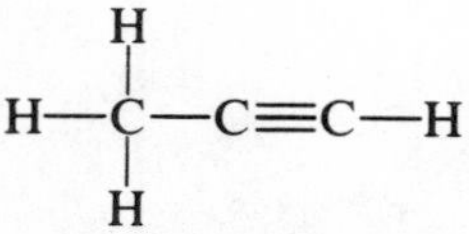

$$H{-}\overset{H}{\underset{H}{C}}{-}C{\equiv}C{-}H \quad \text{or} \quad CH_3{-}C{\equiv}CH \quad \text{or} \quad CH_3C{:}CH$$

propyne (methylacetylene), C_3H_4

and there are two isomeric butynes, C_4H_6.

$$CH_3CH_2C{\equiv}CH \qquad CH_3C{\equiv}CCH_3$$

but-1-yne but-2-yne

The formation of the triple bond by *sp* hybridization and overlap of p_y and p_z orbitals was considered in Chapter 1 (page 8). Since ethyne must be a linear molecule, the question of geometrical isomerism cannot arise. In any case it will be recalled that there is no appreciable restriction to rotation about triple bonds.

The aromatic framework

A very important family of compounds contain this framework, the parent of which is the hydrocarbon benzene, C_6H_6. They are called *aromatic* compounds to distinguish them from the others discussed above which are often collectively known as *aliphatic* compounds (or *alicyclic* compounds if they contain rings). Benzene is a six-membered ring, and it follows from its molecular formula that it must contain the equivalent of three double bonds. Kekulé first wrote the formula as follows in 1865. This formula is often conventionally abbreviated by missing out the CH groups. Systems like this containing alternate single and double bonds are called *conjugated* systems.

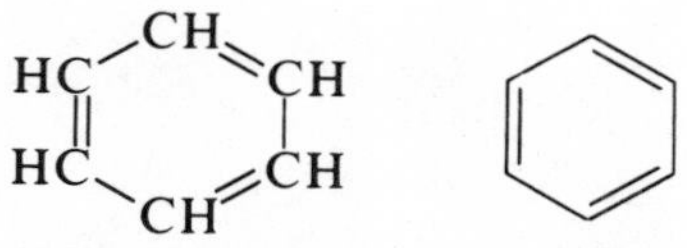

CH HC CH HC CH CH

benzene (Kekulé)

Modern valency theory interprets the benzene molecule as a flat (planar) ring of six sp^2 hybridized carbon atoms. The three sp^2 σ bonds formed by each carbon are thus accounted for and the molecule is a regular hexagon as shown in plan in Figure 3.1.

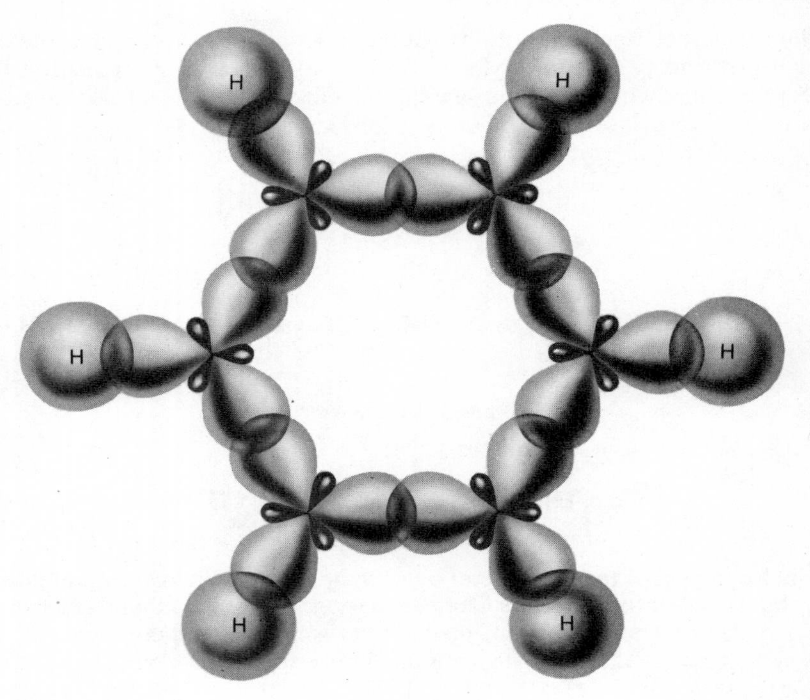

Figure 3.1

However, each carbon atom also has an electron in a p_z orbital, as in Figure 3.2 (a).

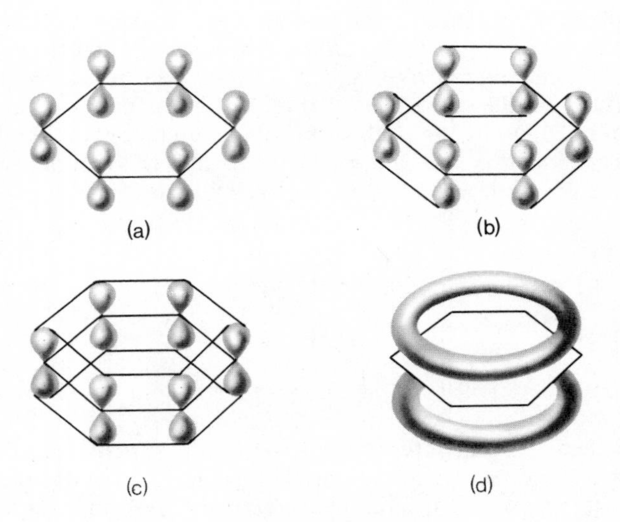

Figure 3.2

Overlap of each of these with *one* neighbour (Figure 3.2 (b)) would give the Kekulé structure. However, overlap of each p_z orbital with *both* its neighbours is obviously possible (Figure 3.2 (c)). The result is that the six p_z electrons are completely delocalized as π electrons in a molecular π orbital covering the whole molecule. The effect is that of thick doughnut-like rings (or *annuli*) of electron density (negative charge) above and below the plane of the ring (Figure 3.2 (d)). There are two notable consequences of this.

1. Because of the gain in stability, or loss in energy, arising from the delocalization of the electrons, the aromatic structure is much more stable than it would be if the three double bonds were localized. The energy associated with this gain in stability, the so-called *delocalization energy* or *resonance energy*, can be calculated from thermochemical data. It is quite large, about 144 kJ mol^{-1}. (This is the value usually quoted, but other values are sometimes given if the comparisons are made on different bases.)

2. The six bonds are now all the same, neither single nor double, but half way between. They are said to have a bond order of 1.5. The C–C bond length in benzene (0.140 nm) is between the lengths of the single (0.154 nm) and double (0.133 nm) carbon–carbon bonds. This is in accordance with the outstanding feature of aromatic chemistry, namely that aromatic compounds do not usually undergo the reactions (especially addition reactions) which are typical of other unsaturated frameworks like alkenes.

Resonance

There is another way of looking at the problem of bonding in aromatic compounds which leads to essentially the same conclusions as the 'molecular-orbital' (M.O.) approach described above. It is sometimes useful to think in this way; the ideas involved are general, and can be applied to other systems.

There are obviously two Kekulé forms of benzene.

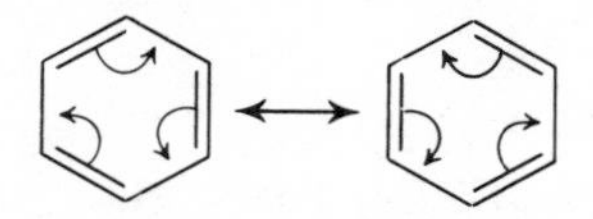

Since all the bonds are in fact equal, and of order 1.5, the actual structure can be thought of as a hybrid of these two structures. It is called a *resonance hybrid* of these *canonical structures*. This simple concept can be put on a mathematical basis using wave mechanical techniques, and this leads to the expected conclusion that the resonance leads to a reduction in energy. This reduction, the resonance energy, comes out of the calculations at about 150 kJ mol^{-1}, very close to the experimental value given above.

The condition for resonance is that the canonical structures must have the same σ bond framework and must be interconvertible simply by movement of electrons, in this case electron pairs, as indicated by the curved arrows. The participation of canonical structures in a resonance hybrid is indicated by the double-headed arrow. It is important to realize that resonance does *not* involve rapid interconversion of molecules in the two forms, so that benzene (in this instance) is *not* a 50–50 equilibrium mixture of molecules in the two forms—this is quite erroneous, as is immediately obvious when one realizes that such a situation could not possibly lead to a diminution in the energy of the system. The actual situation is that *every molecule is the same*, and may be thought of as a *hybrid of* (i.e. *intermediate between*) *the two Kekulé structures*. These canonical structures therefore have no physical reality—only the hybrid is real—but instead are simply mental pictures that are helpful in enabling us to visualize the hybrid structure. For this reason they are sometimes known as 'intellectual scaffolding'.

In fact, the two Kekulé structures contribute about 80 per cent to the resonance hybrid. The other 20 per cent is accounted for by the contributions of the three so-called 'Dewar' structures, which involve long (and therefore very weak) bonds between opposite carbon atoms.

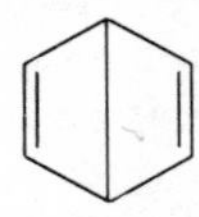 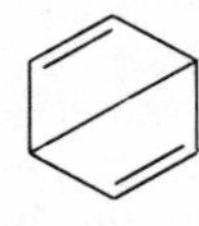 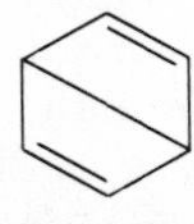

For most qualitative purposes, however, these can be neglected, and benzene can be considered to be a resonance hybrid of the two Kekulé structures. Benzene is conventionally represented either by one of the Kekulé structures or by

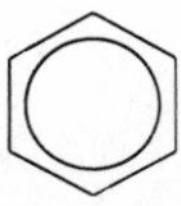

Both representations are useful in different contexts, and both will be used when appropriate in this book.

The resonance approach is sometimes known as the *valence bond* (V.B.) method, to distinguish it from the M.O. method.

Structure determination

The first stage in trying to find out the exact structural nature of an unknown compound is to determine what elements it contains. One can usually assume the presence of carbon and hydrogen (and often oxygen). The other common elements are the halogens, nitrogen, and sulphur, and sometimes metals which are easily identified by the reactions of their ions.

Qualitative analysis

The classic experimental method used to identify these elements is the sodium fusion method, first developed by Lassaigne. A small quantity of the compound is fused with a small pellet of sodium in an ignition tube, and then heated strongly. The black residue is ground up with water, the resulting mixture filtered, and tests carried out on the filtrate. Nitrogen is converted into sodium cyanide, halogens into sodium halides, and sulphur into sodium sulphide. These are present in the filtrate, which is alkaline because of the reaction of the excess of sodium with water to give hydroxide ions. A portion of the filtrate is first tested for cyanide ions by adding a solution of iron(II) sulphate. Green iron(II) hydroxide is precipitated, and the mixture is boiled for a few seconds so that a little of this iron(II) hydroxide is oxidized to iron(III) hydroxide. The mixture is then acidfied, when the hydroxides dissolve. If cyanide ions are present they are converted by the Fe^{2+} ions into hexacyanoferrate(II) (ferrocyanide) ions $[Fe(CN)_6{}^{4-}]$ which, in the presence of the Fe^{3+} ions formed by oxidation, give the familiar Prussian blue. The appearance of a Prussian blue colour or precipitate is therefore indicative of the presence of nitrogen in the organic compound.

Sulphur is present as sulphide ions, and these are easily recognized because they give a black precipitate of lead sulphide with lead ethanoate, or a purple colour with sodium nitroprusside.

Halide ions are tested for with silver nitrate, but if nitrogen and/or sulphur are present the cyanide and/or sulphide ions must first be removed by boiling the filtrate with a little concentrated nitric acid. This is necessary because silver cyanide and silver sulphide are both insoluble.

The Lassaigne procedure can be hazardous, so the following alternative (Middleton's method) is recommended instead. The substance is heated strongly with a mixture of zinc dust and anhydrous sodium carbonate, and the residue ground up with water and filtered as before. Nitrogen and halogens are converted into sodium cyanide and sodium halides respectively, and sulphur into zinc sulphide. Before testing for nitrogen as above, sodium hydroxide solution must be added to the filtrate. Tests for halogens can be done as described for the Lassaigne procedure. To test for sulphur, the *residue* (containing insoluble zinc sulphide) is treated with dilute hydrochloric acid, and the hydrogen sulphide evolved is identified by smell or by the production of a brown stain (lead sulphide) on a filter paper moistened with sodium plumbate(II) (plumbite) solution.

Quantitative analysis

The next stage is to analyze the compound quantitatively for the elements found to be present. The methods are gravimetric and volumetric, but it is not necessary to detail them here. Suffice it to say that with microanalytical techniques only very small quantities (3–5 mg) are required, but of course the compound must have been purified rigorously before analysis. Microanalysis is a job for the specialist, and it is therefore common to use commercial microanalytical laboratories, although the development in recent years of instruments in which the whole process is done automatically has made it easier for even quite small laboratories to conduct their own microanalyses.

An *empirical* (simplest) formula, which indicates the relative numbers of each kind of atom in the molecule, can be calculated from the analytical results, as illustrated by the following example.

An organic compound containing only carbon, hydrogen, and oxygen was found to contain 39.9 per cent of carbon and 6.7 per cent of hydrogen (by weight, of course). It is somewhat more difficult (though now quite possible) to analyze for oxygen, so its percentage is often obtained by difference

(100 − 39.9 − 6.7 = 53.4 per cent). The calculation is then most conveniently set out in tabular form as follows.

Element	*Per cent*	*Relative atomic mass*	*Atomic ratio*	*Simplest ratio*
C	39.9	12	39.9/12 = 3.33	1
H	6.7	1	6.7/1 = 6.7	2
O	53.4	16	53.4/16 = 3.34	1

Empirical formula is CH_2O.

Determination of relative molecular mass

The next stage is to measure the relative molecular mass by the standard physical methods (e.g. vapour density, ebullioscopic, and cryoscopic methods). Mass spectrometry is now increasingly used. This enables the *molecular formula*, which must be the same as or a multiple of the empirical formula, to be calculated. Thus if the relative molecular mass of the compound in the previous example was 180, the molecular formula was $C_6H_{12}O_6$ [$(CH_2O)_6$, because CH_2O = 30].

Structure

The determination of structure once the molecular formula is known can be a long and complex investigation. An examination of the reactions of the compound can lead to a knowledge of the functional groups present, and measurements of the various kinds of spectra of the substance are also extremely valuable in giving structural information.

Infrared spectra and nuclear magnetic resonance (NMR) spectra are particularly useful. Many groups show absorptions in characteristic regions of the infrared spectrum (e.g. $\rangle C{=}C{=}O$ at a frequency of about 1700 cm^{-1}), and their exact position can also give information about the molecular environment of the functional groups. NMR spectra give absorptions due to the hydrogen atoms of the molecule, and characteristic of all the various structurally different hydrogen atoms. Thus, for example, the NMR spectrum of ethanol, CH_3CH_2OH, shows three groups of absorptions corresponding to the hydrogen atoms in the CH_3, CH_2, and OH groups respectively. Their intensities are in the ratio 3:2:1, showing that there are three hydrogen atoms of the first kind, two of the second, and one of the third. Information of this kind obviously makes NMR spectroscopy a most valuable tool in structure determination.

The more classical method of degradation, whereby a complex molecule is chemically broken down into simpler molecules which are then identified, is of course still used. Perhaps its more modern manifestation is mass spectrometry, whereby fragmentation of the molecule occurs in the mass spectrometer, and the relative molecular masses and relative abundances of the fragments are obtained from the mass spectrum. Structural information can then be deduced from these results.

The classic final demonstration of the correctness of a postulated structure is to synthesize it by unambiguous routes such that the final product can have only the structure postulated, and no other. If this product can be shown to be identical with the compound under investigation, the postulated structure must have been correct.

This demonstration of identity is done by comparison of physical properties, particularly spectra. It may be done very simply for solids by showing that they have the same melting point, and that this melting point is not depressed when the two are mixed intimately (ground finely) together; if they were different the melting point would, in the vast majority of cases, be depressed.

Only a very brief outline of this vast and complex activity of structure determination can be given here. The purpose of this section has been only to give the reader some idea of the method of approach to the problem, and of the techniques available for its solution.

Questions

1. Classify and give examples of the various types of isomerism which occur among organic compounds.

(SUJB 'S' paper, 1972)

2. Give reasons for the items which are underlined in the following account of a laboratory test for the detection of nitrogen in an organic compound.

A small amount of the substance is mixed with a small pellet of sodium metal in a small, dry hard glass test tube. The mixture is heated gradually at first and then as strongly as possible until the glass begins to melt. The hot tube is then dropped into a mortar half filled with pure water. Precautions are necessary to avoid injury during this operation.
The mixture is stirred and filtered. To a portion of the filtrate two drops of freshly-prepared iron(II) sulphate solution are added. A greenish precipitate forms. The mixture is then boiled for about one minute, cooled, and a drop of aqueous iron(III) chloride added. If, on the addition of an excess of hydrochloric acid, a bluish-green coloration or a blue precipitate is formed, the original compound contained carbon and nitrogen.

If the above substance also contained sulphur, what further test would you carry out to prove this? Explain the reactions in the test for sulphur.

(JMB, 1971)

Chapter 4

Classification of Reagents and Reactions

Homolysis and heterolysis

It has already been stated that the cleavage of a covalent bond is the fundamental process of organic reactions. It is now necessary to examine in some more detail the ways in which this can take place. If we consider a molecule AB in which the groups A and B are joined by a single covalent bond, there are obviously two types of cleavage. The first type (1) in which the electron pair remains intact and becomes the property of either A or B (depending on which

$$A{:}B \longrightarrow A^+ + {:}B^- \quad \text{heterolysis} \qquad (1)$$

$$A{:}B \longrightarrow A\cdot + \cdot B \quad \text{homolysis} \qquad (2)$$

is the more electronegative) is called *heterolytic fission* or *heterolysis*, and if AB is neutral gives rise to the formation of a pair of oppositely charged ions. The other type (2) is called *homolytic fission* or *homolysis* and gives rise to uncharged particles (atoms or *free radicals*), each of which contains an odd electron with an unpaired spin.

At first sight, it might be thought that since heterolysis involves the separation of oppositely charged particles which mutually attract one another, it should normally require more energy than homolysis, in which the particles to be separated are both neutral. If this were the case, homolysis would be expected to be much the more common, and heterolysis quite exceptional. This is, of course, not so; both types of process are common, and indeed heterolytic reactions are probably the more familiar. The clue to this apparent paradox is that in the gas phase, where the molecules are relatively isolated from their environment, homolysis is much more common, and so the above argument, which ignores the environment, has more validity. It is the influence of the environment (the solvent) which makes heterolysis a viable process in the liquid phase because the increment in energy required to overcome the electrostatic (or 'Coulombic') attraction of the ions is compensated by the decrease in energy resulting from the electrostatic association of the ions with the solvent molecules (solvation). The better the solvating properties of the solvent, or in physical terms the higher its dielectric constant, the more it will tend to favour heterolysis. Heterolysis will also be more likely if the molecule AB is extensively polarized. For example haloalkanes, which are polarized $\overset{\delta+}{R}$—$\overset{\delta-}{X}$ by virtue of the $+I$ effects of the alkyl groups and the $-I$ effects of the halogens so that there is an excess of electron density at the halogen and a deficit at the alkyl group, would be expected to undergo heterolysis readily.

Homolysis on the other hand can occur at reasonable rates at temperatures which are convenient in the laboratory if the dissociation energy, *D* (A–B), is of the order of 80 – 160 kJ mol^{-1} usually provided that the following conditions, which are the opposites of those favouring heterolysis, are satisfied.

(1) The molecule should not be extensively polarized, but should be electrically fairly symmetrical.

(2) The solvent should be a poorly solvating one having a low dielectric constant (e.g. alkanes, benzene, tetrachloromethane), rather than the good solvating solvents such as water and alcohols, which favour heterolysis.

Electrophiles and nucleophiles

The fragments resulting from heterolysis can themselves be regarded as examples of the two types of heterolytic reagent which can attack molecules, forming new bonds. The fragment A^+ is electron-poor, having a vacant orbital in its valency shell which can accommodate an electron pair. It must therefore form a new bond by accepting a share in an electron-pair donated by the molecule which it attacks. Thus it may, for example, attack an alkene forming a new bond, the electron pair of which is the π electron pair of the double bond (3).

The movement of the electron pair is conventionally indicated by a curved

$$\text{>C=C<} + A^+ \longrightarrow \text{>C(A)}\text{—}\overset{+}{\text{C}}\text{<} \qquad (3)$$

arrow in this and many subsequent diagrams. The tail of the arrow indicates the origin of the electron pair and its head their destination.

Because of their electron-poor nature, reagents of this kind will most readily attack centres of high electron density, like alkenes or *arenes* (aromatic compounds), on account of the cloud of π electron density surrounding such molecules. Entities like A^+ are therefore known as *electrophilic* (electron-loving) reagents, or *electrophiles*. They are usually but not invariably cations, and common examples in organic chemistry are H^+, Br^+, and NO_2^+.

The fragment $:B^-$ in (1) has the opposite properties, and is known as a *nucleophilic* (nucleus-loving) reagent, or *nucleophile*. Nucleophiles have one or more unshared pairs of electrons which they can donate to form a new bond with the atom which they attack. They therefore tend to react more readily at electron-poor (positive) centres. They can, for example, attack saturated carbon atoms, particularly when they are attached to electron-attracting groups, such as halogens, which tend to withdraw electron density from them (4). Again the movement of the electron-pairs is indicated by the curved arrows.

$$\ddot{B}^- + R\text{—}X \longrightarrow B\text{—}R + \ddot{X}^- \qquad (4)$$

All nucleophiles possess unshared electrons. While many are anions (e.g.

HO^-, Hal^-, CN^-), there are a number of important nucleophiles which are electrically neutral (e.g. H_2O, NH_3, NR_3, SR_2). What they have in common, however, is that when they react, they always become one unit less negatively charged [see (4) above, B^- becoming B in B–R], or one unit more positively charged, which amounts to the same thing. Electrophiles on the other hand obey the opposite rule, becoming one unit *less* positively charged (or one unit *more* negatively charged) on reaction.

Free radicals

The uncharged fragments resulting from homolysis (2) are atoms or free radicals. The definition of a free radical is that it should possess an unpaired electron. This definition embraces some familiar stable molecules, notably NO and NO_2, which are in fact free radicals and exhibit many of the physical and chemical properties which characterize these entities.

Free radicals are often characterized by very high reactivity, which arises from the large decrease in energy (or increase in stability) resulting from electron pairing to form a covalent bond. Being highly reactive, they tend to be less discriminating (less 'choosy') about the sites at which they will react than electrophiles or nucleophiles, so free radical reactions sometimes tend to be rather complex and to lead to mixtures of products.

Since free radicals do not carry formal charges, they do not have pronounced polar preferences with regard to the sites they attack, although small polar preferences arising from the electronegativity of the atom carrying the unpaired electron can sometimes be observed. Free radicals are therefore capable of attacking molecules of almost all types, both electron-rich and electron-poor, and hence undergo a very wide variety of reactions. This versatility is very valuable and consequently free radical reactions are widely used and are of great importance in many industrial processes.

Homolytic reactions always involve 'unpairing' of the electron pairs constituting covalent bonds, and the new bonds are formed by donation of one electron from each of the two entities forming the bonds. It is single electrons therefore, rather than pairs, which move about the molecules in homolytic reactions. This movement of single electrons is represented by 'half-headed' curved arrows as in (5). This equation illustrates a very common homolytic process, that of hydrogen-abstraction in which a radical attacks a hydrogen atom attached to a saturated carbon framework to give another radical.

$$\dot{A} + H—R \longrightarrow A—H + R\cdot \qquad (5)$$

Examples of this process are discussed in Chapter 5 (pages 41, 43).

Classification of reactions

The wide variety of organic reactions which exist fall into four main classes, namely substitution, addition, elimination, and rearrangement reactions. Examples of the first, second, and third types will be met frequently, but most rearrangements are beyond the scope of an elementary course.

However, it will be useful to explain and exemplify the four types here.

Substitution

In these reactions one atom or group in a molecule is replaced by another. Reaction (4) (page 33) is an example of a nucleophilic substitution, and nitration of benzene, which is discussed in detail in Chapter 7 (pages 61–6), is an example of an electrophilic substitution (6). In this reaction the attacking reagent is the electrophile NO_2^+.

$$NO_2^+ + C_6H_6 \longrightarrow C_6H_5NO_2 + H^+ \tag{6}$$

Free radicals also take part in substitution reactions, of which (5) is an example. Here the substitution is at hydrogen rather than carbon, and the replaced group is R. Substitution is sometimes referred to as 'replacement' or 'displacement'.

Addition

A reagent XY can *add* to unsaturated compounds of many types. For example, compounds containing carbon–carbon double bonds can undergo addition reactions of the type of (7) to give saturated products.

$$\rangle C{=}C\langle + XY \longrightarrow \rangle C(X){-}C(Y)\langle \tag{7}$$

Examples of electrophilic, nucleophilic, and homolytic additions will be met from time to time (e.g. Chapter 6, pages 47–54; Chapter 10, pages 106–8).

Elimination

This is essentially the reverse of addition, an unsaturated compound being formed by loss of the two constituents of a (usually) small molecule from the two atoms concerned. An example is the formation of an alkene by elimination of hydrogen bromide from a bromoalkane, usually under the influence of a base (cf. Chapter 8, page 82).

$$\rangle CH{-}CBr\langle \xrightarrow[(-HBr)]{\text{base}} \rangle C{=}C\langle \tag{8}$$

Rearrangement

As the name implies, these reactions involve the rearrangement of the molecular skeleton by migration of atoms or groups from one atom to another within the molecule. These reactions can be quite complex and their mechanisms, which have traditionally fascinated chemists, have often proved of great interest.

Questions

1. Most reactions encountered in the chemistry of carbon compounds involve the breaking (and subsequent formation) of covalent bonds.

   Discuss the ways in which a covalent bond can be broken during a reaction and examine the influence of different types of reagents on the mechanism of bond breaking in carbon compounds.

   (London, Special Paper, 1973)

2. Draw up a list showing how organic reagents are classified and apply this classification to a mechanistic consideration of reactions of simpler type.

   (SUJB, 'S' paper, 1973)

Chapter 5

Alkanes

Nomenclature

Some simple alkanes are given common or 'trivial' names, often using the prefixes n- to indicate a straight chain and iso- to indicate a chain branched so that it contains the group $(CH_3)_2CH-$. The following examples illustrate this:

$CH_3CH_2CH_2CH_3$	$(CH_3)_2CHCH_3$	$(CH_3)_2CHCH_2CH_2CH_3$
n-butane	isobutane	isohexane

While such names are widely used for simple compounds, a more systematic method is required for more complex structures, and rules devised by the International Union of Pure and Applied Chemistry (I.U.P.A.C.) are accepted internationally. In this country, recommendations for the naming of chemical compounds based on the I.U.P.A.C. rules have recently been devised by the Association for Science Education (A.S.E.)* and have gained wide acceptance in schools. These 'ASE' names are given in this book, but for those compounds for which alternative systematic or trivial names are currently in common use, these alternatives are also given.

The fundamental feature of the systematic naming of organic compounds is that a compound is named as a derivative of the longest unbranched chain it contains, and for alkanes the suffix -ane is used. For chains of up to four carbon atoms the names methane, ethane, propane, and butane are used, and from five carbons upwards Latin or Greek numerals indicate the number of carbon atoms in the chain, e.g.

C_5	pentane	C_9	nonane
C_6	hexane	C_{10}	decane
C_7	heptane	C_{11}	undecane
C_8	octane	C_{12}	dodecane, etc.

The carbon atoms in the longest chain are numbered consecutively from the end which gives the lower numbers to the atoms bearing substituents. Thus on this system isobutane becomes 2-methylpropane and isohexane becomes 2-methylpentane. The following examples also illustrate the method.

$$\overset{1}{C}H_3\overset{2}{C}(CH_3)_2\overset{3}{C}H_3$$

$$\overset{1}{C}H_3\overset{2}{C}H(CH_3)\overset{3}{C}H(CH_2CH_3)\overset{4}{C}H_2\overset{5}{C}H_2\overset{6}{C}H_2\overset{7}{C}H_3$$

2-methyl-3-ethylheptane

2,2-dimethylpropane (the trivial name 'neopentane' is often used for this compound)

* *Chemical Nomenclature, Symbols and Terminology for Use in School Science* (Hatfield: Association for Science Education, 1972)

$$\overset{1}{C}H_3\overset{2}{C}H_2\overset{3}{C}H(CH_3)\overset{4}{C}H(CH_3)\overset{5}{C}H_2\overset{6}{C}H_3$$

(drawn with H_3C and CH_3 on carbons 3 and 4) [or $\overset{1}{C}H_3\overset{2}{C}H_2\overset{3}{C}H(CH_3)\overset{4}{C}H(CH_3)\overset{5}{C}H_2\overset{6}{C}H_3$]

3,4-dimethylhexane (note alternative way of writing structure, which is more convenient to print)

$$CH_3\overset{3}{C}H(\overset{2}{C}H_2\overset{1}{C}H_3)-\overset{4}{C}H(CH_2CH_3)\overset{5}{C}H_2\overset{6}{C}H_2\overset{7}{C}H_3$$

3-methyl-4-ethylheptane

The last example illustrates the point that the longest chain is not always immediately obvious, and some care is necessary in looking for it.

Similar rules are applied to the naming of 'alkyl groups', which are the monovalent groups derived by removal of a hydrogen atom from an alkane. The suffix -yl is now used, with numerals to indicate the position of the hydrogen which has been removed; e.g.

$CH_3CH_2—$	$CH_3CH_2CH_2—$	$CH_3\underset{\vert}{C}HCH_3$
ethyl	prop-1-yl	prop-2-yl

The trivial names n-propyl and isopropyl are still quite commonly used for prop-1-yl and prop-2-yl, respectively.

Occurrence

Alkanes occur naturally as natural gas, which is largely methane together with smaller quantities of the other gaseous alkanes, and as petroleum. The composition of petroleum varies with its source, but it can contain straight

Table 5.1

B.p./°C	*Approximate composition*	
<80	C_5–C_7	Light petroleum, solvent
80–200	C_6–C_{11}	Petrol, motor fuel
200–300	C_{12}–C_{16}	Paraffin (kerosene), lighting and heating
>300	C_{13}–C_{18}	Fuel oil
	$>C_{18}$	Lubricating oils and greases, vaseline, paraffin wax for candles etc.

and branched chain alkanes up to about C_{40}, together with cycloalkanes (known in the petroleum industry as naphthenes), aromatic hydrocarbons, and impurities. Petroleum is first 'cleaned up' to remove dirt, sand etc., and then dried. It can then be distilled into fractions, each of which is a mixture of hydrocarbons (see Table 5.1).

Cracking

Since there is a great demand for the lighter (more volatile) fractions, particularly the petrol fraction and the gaseous saturated and unsaturated lower members which are the basis of the petrochemicals industry, it is advantageous to break down ('crack') the larger molecules in the heavier fractions. This cracking is done essentially by heating the material to high temperatures (500–600 °C) when rather indiscriminate homolytic fission of many bonds occurs, and mixtures of simpler saturated and unsaturated hydrocarbons and some hydrogen are formed, e.g.

$$C_nH_{2n+2} \longrightarrow C_xH_{2x+2} + C_yH_{2y}$$

where $x + y = n$.

Cracking of heavy oils is done in the liquid phase at high pressures, and cracking of less heavy fractions in the gas phase at much lower pressures. Catalysts consisting of mixtures of oxides of elements such as silicon, aluminum, and thorium are also used. The simple alkenes ethene ($CH_2{=}CH_2$) and propene ($CH_3CH{=}CH_2$) (see Chapter 6) are major products of petroleum cracking.

Knocking and anti-knock agents

The phenomenon of 'knocking' in petrol engines arises from premature ignition of the fuel-air mixture. It is wasteful of energy, and therefore to be avoided. Some hydrocarbons (e.g. highly branched alkanes, aromatics) are better from this point of view than others. The C_8 hydrocarbon 2,2,4-trimethylpentane ('iso-octane') is particularly good, and is given an 'octane number' of 100 on an arbitrary scale.

$$(CH_3)_3C.CH_2CH(CH_3)_2$$

'iso-octane'

The octane number of a fuel on this scale is a measure of its anti-knock properties. Octane numbers can be raised by certain additives, notably tetraethyl-lead (TEL), $(C_2H_5)_4Pb$, and are lowered by sulphur-containing impurities. Most commercial motor fuels currently available therefore contain TEL, and sulphur compounds are carefully removed during the refining of the petroleum. However, addition of TEL causes environmental problems because of the toxicity of lead compounds, and requires an additional additive in the petrol to remove lead from the engine.

Other sources

Petroleum and other fossil fuels were probably formed by anaerobic fermentation (i.e. fermentation in the absence of air) of vegetable matter. Marsh gas, which is formed by anaerobic fermentation of cellulose, consists mainly of methane.

Shale is a rich source of alkanes, and coal is largely an elaborate network of carbon rings. Alkanes can be obtained by the low temperature distillation of coal, and have also been prepared on an industrial scale for use as fuels by the catalytic hydrogenation of coal.

Mixtures of alkanes for use as fuels can also be synthesized from water gas (or 'synthesis gas', a mixture of carbon monoxide and hydrogen) by passing it at 200–300 °C under pressure over various catalysts (metals and metal oxides). This is the *Fischer–Tropsch process*.

Although these sources have been more expensive than petroleum, there is no doubt that they will increase in importance as world resources of petroleum become depleted and its price increases.

It is not usually necessary to prepare alkanes as such in the laboratory, but there are of course many reactions in which they are formed. Some of these are important and some are of more general application. The more important ones will be met in various contexts, and mention of these is made here with page references for the reader's convenience. They are: hydrogenation of alkenes (page 47), from halogenoalkanes (alkyl halides) by reduction (page 76) and by the Wurtz reaction (page 76), from carboxylic acid salts by decarboxylation (page 124), and hydrolysis of Grignard reagents (page 77).

Properties and reactions of alkanes

The liquid alkanes are oily substances. They are all insoluble in water, but soluble in organic solvents.

Since they are electrically symmetrical, essentially non-polar substances, and are insoluble, generally, in solvents of high dielectric constant (i.e. good ionizing or solvating solvents, or polar solvents, e.g. water), it could be predicted that alkanes should generally undergo homolytic rather than heterolytic reactions. This is in fact the case, and the more important of these reactions will now be discussed.

Halogenation

Methane combines with chlorine in ultraviolet light (e.g. in bright sunlight).

$$CH_4 + Cl_2 \longrightarrow HCl + CH_3Cl$$

chloromethane
(methyl chloride)

$$CH_3Cl + Cl_2 \longrightarrow HCl + CH_2Cl_2$$

dichloromethane
(methylene chloride)

$$CH_2Cl_2 + Cl_2 \longrightarrow HCl + \underset{\text{trichloromethane (chloroform)}}{CHCl_3}$$

$$CHCl_3 + Cl_2 \longrightarrow HCl + \underset{\text{tetrachloromethane (carbon tetrachloride)}}{CCl_4}$$

The reaction may become explosive, and usually it gives a mixture of products. No reaction takes place in the dark. With higher alkanes any or all of the hydrogens can be replaced by chlorine and the mixture of products can become highly complex.

For the reasons given above, these reactions are homolytic, as indeed are virtually all light-catalyzed (*photolytic*) reactions. The function of the light is of course to supply quanta of energy to cleave bonds (in this case the Cl–Cl bond), but it is by no means necessary that there should be one quantum for every molecule which reacts. Indeed the number of molecules which react for each quantum supplied (the *quantum yield*) can be very large, because these are usually *chain reactions*. Many, but not all, homolytic processes have chain mechanisms.

The mechanism of the chlorination of methane is as follows, and those of chlorination of any hydrocarbon RH to give RCl, and of the later stages of the chlorination of methane to give dichloromethane etc., are analogous to this mechanism.

Initiation

$$Cl_2 \xrightarrow{h\nu} 2Cl\cdot \qquad \Delta H° = 244 \text{ kJ mol}^{-1}$$

Propagation

$$Cl\cdot + CH_4 \longrightarrow HCl + CH_3\cdot \qquad \Delta H° = -4 \text{ kJ mol}^{-1}$$

$$CH_3\cdot + Cl_2 \longrightarrow CH_3Cl + Cl\cdot \qquad \Delta H° = -101 \text{ kJ mol}^{-1}$$

etc.

Termination

$$2CH_3\cdot \longrightarrow C_2H_6$$

$$CH_3\cdot + Cl\cdot \longrightarrow CH_3Cl$$

$$2Cl\cdot \longrightarrow Cl_2$$

The following points are noteworthy.

(1) There are always these three stages in any chain mechanism. If the chains are reasonably long, the bulk of the products are formed by the propagation reactions: termination products are formed only in small quantities because only one molecule is formed for every chain terminated. In this case

the chains are very long (about a million stages) so that only one molecule of a termination product is formed for every million molecules of chlorination product. There are usually several alternative terminations, as in this case, because there are several radical species present, and they may all combine with one another. Termination reactions of this kind, which involve combination of radicals, are exothermic, the amount of energy released on formation of the new bond being equal to that which would be required to break it. The presence of some 'third body', which can be the walls of the container, is therefore required to accept and dissipate this released energy. Note that in initiation stages radicals are *formed*, in termination stages radicals are *destroyed*, and in propagation stages one radical is *exchanged* for another.

(2) This is a gas-phase reaction and the conditions therefore favour homolysis. Liquid-phase halogenation is usually conducted without a solvent, or in a non-polar solvent so that again the conditions favour homolysis. For example, methylbenzene, $C_6H_5CH_3$, (which is usually known by its trivial name 'toluene') is photolytically chlorinated in the 'side chain' at its boiling point. This is characterictic of such compounds.

$$C_6H_5CH_3 \xrightarrow{Cl_2} C_6H_5CH_2Cl \longrightarrow C_6H_5CHCl_2 \longrightarrow C_6H_5CCl_3$$

Aromatic *nuclei* (benzene rings) do not react with halogens in this way, but undergo other types of reactions. Thus this kind of homolytic chlorination, the essential step of which is the hydrogen-abstraction which is the first of the propagation reactions, is characteristically undergone at sp^3 carbon. It is sp^3 C–H bonds that are susceptible to hydrogen abstraction.

(3) The cleavage of the Cl–Cl bond in the initiation stage is endothermic and requires the supply of 244 kJ mol^{-1} of energy. This is supplied by the light. Visible light is suitable because, for example, the size of the energy quantum, $h\nu$, for light at 400 nm is 300.7 kJ mol^{-1}. The function of the radiation is simply to provide energy in packets of about the right size to cleave the required bonds.

(4) Chains cannot be sustained if any stage in their propagation is appreciably endothermic. Here the reaction proceeds because neither stage is endothermic, but the corresponding reaction with bromine does not work because of the endothermicity of the abstraction stage.

$$Br\cdot + CH_4 \longrightarrow HBr + CH_3\cdot \qquad \Delta H^\circ = +63 \text{ kJ mol}^{-1}$$

Oxidation

The complete oxidation of alkanes by burning in oxygen or air to carbon dioxide and water is the source of energy which gives them their use as fuels.

$$2C_nH_{2n+2} + (3n+1)O_2 \longrightarrow 2nCO_2 + (2n+2)H_2O + \text{heat}$$

The heat of combustion of methane is 885.8 kJ mol^{-1}, and heats of combustion for the members of the homologous series of alkanes increase by about 630 kJ mol^{-1} per CH_2 increment.

More controlled aerial oxidation leads to the conversion of RH into the hydroperoxide, R–O–O–H. This reaction is important because it can lead to the degradation (breakdown) of many plastic materials or other polymers in air, and steps must often be taken to retard it in order to preserve the durability of the material. On the other hand, there may be circumstances in which one might wish to promote it, for example, to allow plastic litter to break down in air.

The term '*autoxidation*' is usually applied to oxidation of organic compounds by atmospheric oxygen. It is a rather poor name because, although oxygen is required, the prefix 'aut-' wrongly implies that the process takes place without external agency. However the term is so widely used that it is necessary to be familiar with it.

The mechanism is again a radical chain initiated by light or an *initiator*, which can be any substance whose thermal breakdown gives radicals (X·) to start the chains. The reactions involved are as follows, and follow the classical pattern for a radical chain process.

Initiation

$$\text{Light or initiators} \longrightarrow X\cdot$$
$$X\cdot + O_2 \longrightarrow XOO\cdot$$
$$XOO\cdot + RH \longrightarrow XOOH + R\cdot$$

Propagation

$$R\cdot + O_2 \longrightarrow ROO\cdot$$
$$ROO\cdot + RH \longrightarrow ROOH + R\cdot \quad \text{etc.}$$

Termination

$$2R\cdot \longrightarrow R_2$$
$$R\cdot + ROO\cdot \longrightarrow ROOR$$
$$2ROO\cdot \longrightarrow ROOR + O_2$$

An autoxidation reaction of great commercial importance is that of (1-methylethyl) benzene (also known as 'isopropylbenzene' or 'cumene') which is obtained from petroleum. Acid hydrolysis of the hydroperoxide so formed

$$\underset{\text{cumene}}{C_6H_5CH(CH_3)_2} \xrightarrow{\text{autoxidation}} C_6H_5C(CH_3)_2OOH \xrightarrow{\text{acid}} \underset{\text{phenol}}{C_6H_5OH} + \underset{\text{acetone}}{(CH_3)_2CO}$$

gives the substances commonly known as phenol and acetone,* which are used in vast quantities. This reaction is now the main source of phenol.

Sulphonation

Treatment of the higher alkanes with fuming sulphuric acid gives *sulphonic acids*.

$$RH + HO.SO_2.OH \longrightarrow \underset{\text{alkanesulphonic acid}}{RSO_2.OH} + H_2O$$

These products are strong monobasic acids. Their salts are used as detergents (see page 126).

Nitration

Treatment of alkanes in the gas phase with nitric acid vapour (which probably contains oxides of nitrogen and other decomposition products) leads to introduction of the *nitro*- group, $-NO_2$, to give nitroalkanes by a homolytic mechanism. Mixtures of nitroalkanes are formed because of replacement of various hydrogen atoms, and because some rearrangement and cleavage of the carbon skeleton occur under these vigorous conditions.

$$RH + HONO_2 \longrightarrow \underset{\text{nitroalkane}}{RNO_2} + H_2O$$

Questions

1. Some hydrocarbons are said to be 'saturated', others 'unsaturated'. Explain what these terms mean, both in terms of reactions and of structure. (Aromatic compounds should not be discussed.)
   Give an account of the reactions of the two isomers of formula C_6H_{12} (cyclohexane and hex-l-ene) to illustrate the difference; indicate the possible mechanisms of the reactions you describe.
   (London, 1974)
2. Under what conditions do (*a*) aromatic hydrocarbons (e.g. benzene) and (*b*) aliphatic hydrocarbons (e.g. methane) react with chlorine to give substitution products?
   Account for these requirements for different conditions in terms of the natures of the hydrocarbons and the reaction mechanisms involved.
   Why does benzene undergo a substitution reaction with chlorine whereas, under the same conditions, ethene (ethylene) undergoes an addition reaction?
   (O. and C., 1975)

* For systematic nomenclature, see page 86 and page 102, respectively.

Chapter 6

Alkenes and Alkynes

Alkenes

Nomenclature

For systematic nomenclature the same rules are applied in naming the skeleton as for the alkanes, but the appropriate suffix is now -ene, and the position of the double bond is indicated by preceding the suffix with the number of the carbon atom of the double bond which is nearer the end of the chain. The following examples illustrate this.

$$CH_2{=}CH_2$$
ethene

$$CH_3CH{=}CH_2$$
propene

$$\overset{4}{C}H_3\overset{3}{C}H_2\overset{2}{C}H{=}\overset{1}{C}H_2$$
but-1-ene

$$\overset{4}{C}H_3\overset{3}{C}H{=}\overset{2}{C}H.\overset{1}{C}H_3$$
but-2-ene

$$\overset{4}{C}H_3.\overset{3}{C}H_2\overset{2}{C}(CH_3){=}\overset{1}{C}H_2$$
2-methylbut-1-ene

$$\overset{4}{C}H_2{=}\overset{3}{C}(CH_3){-}\overset{2}{C}(CH_3){=}\overset{1}{C}H_2$$
2,3-dimethylbuta-1,3-diene

The derived monovalent groups (*alkenyl groups*) have the suffix *-enyl*, e.g. $CH_3CH{=}CH{-}$, prop-1-enyl; $CH_2{=}CH{-}$, ethenyl (but commonly called *vinyl*); $CH_2{=}CH{-}CH_2{-}$, prop-3-enyl (but commonly called *allyl*).

The trivial names ethylene and propylene are used commonly for the first and second members. Alkenes in general are also sometimes called 'olefins', but this term is now used less frequently than in the past.

Preparation

Alkenes are obtained on a large scale by the cracking of petroleum, which has already been discussed. However, laboratory preparation of alkenes is important because the methods are general ones for the introduction of double bonds, and are not confined to the hydrocarbons themselves. It is often necessary to introduce double bonds into frameworks to which other functional groups may already be attached.

The most important general reaction in which double bonds are formed is the elimination reaction (page 35), and it can be used in several ways to form alkenes:

$$\underset{A}{\overset{|}{>C}}{-}\underset{B}{\overset{|}{C<}} \xrightarrow{-AB} {>}C{=}C{<}$$

For example, water can be eliminated from hydroxy-compounds (alcohols).

$$RCH_2CH_2OH \xrightarrow{-H_2O} RCH{=}CH_2$$

This can be accomplished directly by passing the alcohol vapour over heated alumina. Alcohols, and indeed some other series, are classified according to the degree of chain branching at the carbon atom to which the functional group is attached, thus

$R{-}CH_2OH$ primary alcohols (1 ry)

$RR'CHOH$ secondary alcohols (2 ry)

$RR'R''COH$ tertiary alcohols (3 ry)

[The items 'primary', 'secondary', and 'tertiary' are also applied to amines, but here they are used with a rather different meaning (cf. Chapter 12, pages 142–3), and a little care needs to be taken not to confuse the two usages for the terms.]

The ease of dehydration of alcohols is in the order 3 *ry* > 2 *ry* > 1 *ry*, the temperatures required being about 150 °C, 250 °C, and 350 °C respectively.

These dehydrations can also be brought about by the action of an acid catalyst. Thus treatment of ethanol with concentrated sulphuric acid at about 170 °C gives ethene. The acid catalyst is necessary because –OH is very difficult to detach as OH^- from a carbon skeleton (it is a poor 'leaving group'). Under the influence of the acid a proton is added to convert –OH into $-\overset{+}{O}H_2$. The leaving group is then H_2O, which is much more stable and hence more easily detached than OH^-. Thus many reactions of alcohols are acid-catalyzed and proceed in this way. The sequence of events in this reaction is

$$CH_3CH_2OH + H^+ \xrightarrow{H_2SO_4} CH_3CH_2\overset{+}{O}H_2 \longrightarrow H_2O + CH_3CH_2^+$$

$$CH_3CH_2^+ \longrightarrow CH_2{=}CH_2 + H^+$$

The entity $CH_3CH_2^+$ is called a *carbonium ion*, and such ions are intermediate in many heterolytic reactions.

It is worth noting at this stage that when a carbonium ion is formed, the hybridization of the carbon atom changes from sp^3 to sp^2 when it becomes positively charged. The orbital which is unoccupied in the ion is the p_z orbital, and it will be remembered that the sp^2 orbitals are trigonally disposed and lie in a plane (cf. Chapter 1, page 7). Thus carbonium ions (and, indeed, alkyl radicals, to which similar considerations apply) have planar rather than tetrahedral geometry. This has important consequences, which are further discussed in Chapter 11 (page 136).

Hydrogen halides can be eliminated from alkyl halides (haloalkanes) to give alkenes. This reaction is base-catalyzed, and the most common reagent is 'alcoholic potash', i.e. a solution of potassium hydroxide in ethanol, which contains the strong base $C_2H_5O^-$, the ethoxide ion. The process works only very poorly with primary alkyl halides, because they react preferentially in another way (to give ethers, see Chapter 8, page 83). Better yields are however obtained with secondary and tertiary alkyl halides. Thus while treatment of bromoethane with alcoholic potash is not a practicable way of preparing ethene, propene can be prepared in this way from 2-bromopropane.

$$CH_3\underset{\underset{Br}{|}}{C}HCH_3 \xrightarrow{OEt^-} CH_3CH{=}CH_2 + HOEt + Br^-$$

The mechanisms of these eliminations are discussed in more detail in Chapter 8 (pages 82–3).

There are a number of other ways in which double bonds can be introduced into molecules, but they need not be discussed here.

Properties and reactions of alkenes

The physical properties of alkenes are similar to those of alkanes. The lower homologues (up to the butenes) are gases at room temperature; pent-1-ene boils at 30 °C.

The reactions undergone by alkenes are chiefly those of the double bond, and are generally *addition* reactions (cf. Chapter 4, page 35).

$$>C{=}C< + AB \longrightarrow >\underset{\underset{A}{|}}{C}-\underset{\underset{B}{|}}{C}<$$

They can proceed by both heterolytic (electrophilic) and homolytic mechanisms.

Hydrogenation

Alkenes combine with gaseous hydrogen in the presence of catalysts to give alkanes.

$$>C{=}C< + H_2 \longrightarrow >CH-CH<$$

Many catalysts have been used, and the temperature employed depends on the catalyst. Thus finely divided platinum, palladium, and nickel ('Raney' nickel, prepared by removing the aluminium from a Ni/Al alloy with caustic soda) are effective at room temperature. Nickel on a support such as asbestos is used industrially at 200–300 °C. This is the *Sabatier–Senderens* method, and is used in hydrogenating the double bonds present in vegetable oils to give solid fats in the manufacture of margarine. The process is known as 'hardening'.

The mechanism of catalytic hydrogenation is not fully understood, but it is considered that both alkene and hydrogen are adsorbed on the catalyst surface (hence the requirement for finely divided catalysts with large surface areas). The hydrogen acts as if it were present as hydrogen atoms, i.e. the process is essentially homolytic.

Addition of halogens

Chlorine and bromine add readily to alkenes. If ethene is passed into bromine at room temperature, 1,2-dibromoethane is formed. The decolourization of a solution of bromine in tetrachloromethane in the cold is considered a test for a double bond.

$$CH_2{=}CH_2 + Br_2 \longrightarrow CH_2Br.CH_2Br$$

Halogen addition is usually an electrophilic process, although a homolytic mechanism can occur under appropriate conditions. The key to the electrophilic mechanism of addition is the electron-rich nature of the double bond, which, it will be remembered, is swathed in π electron density above and below the plane of the molecule.

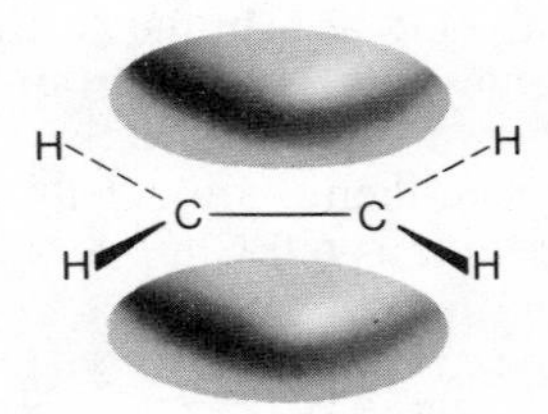

When such a molecule is approached by a bromine molecule, the high electron density of the alkene induces polarization of the bromine molecule and encourages its heterolysis in the sense that Br^+ adds to the alkene, leaving Br^-. It can be demonstrated by investigation of suitable systems that addition of the two atoms of bromine characteristically takes place on opposite sides of the alkene molecule ('*trans*-addition'). This is consistent with an interpretation in which the bromine which first adds to the alkene is attached to *both* carbon atoms of the double bond, the intermediate being formulated as a '*bridged bromonium ion*'. The electron pairs move to form these new bonds as follows.

$$\mathrm{H_2C{=}CH_2} \;+\; :\ddot{\mathrm{Br}}:^{\delta+}\!-\!:\ddot{\mathrm{Br}}:^{\delta-} \longrightarrow \mathrm{H_2C{-}CH_2}\ (\text{with } :\ddot{\mathrm{Br}}:^{+} \text{ bonded to both C}) \;+\; \mathrm{Br}^-$$

Bromide ions then attack this intermediate from the '*rear*' (because the 'front' is shielded by the bulky bromine atom rigidly attached to both carbon atoms).

$$\text{(bromonium ion } H_2C\text{—}CH_2 \text{ bridged by } Br^+ \text{, attacked by } Br^-) \longrightarrow H\text{—}\underset{H}{\overset{Br}{C}}\text{—}\underset{Br}{\overset{H}{C}}\text{—}H$$

In the presence of water, an alternative second stage is possible, in which a water molecule adds in place of the bromide ion.

$$H_2O{:} + H_2C\text{—}CH_2 \,(\text{bridged by } Br^+) \longrightarrow \overset{H_2\overset{+}{O}}{CH_2}\text{—}\underset{Br}{CH_2} \longrightarrow CH_2(OH)CH_2Br + H^+$$

2-bromoethanol
(ethylene bromohydrin)

These *halohydrins* readily lose halogen acids to give *epoxides* (alkene oxides).

$$CH_2(OH)CH_2Cl \xrightarrow{-HCl} H_2C\text{—}CH_2 \,(\text{bridged by } O)$$

epoxyethane
(ethylene oxide)

Addition of hydrogen halides

Hydrogen halides, HX, add readily to alkenes, hydrogen iodide being the most reactive and hydrogen bromide and hydrogen chloride less so; e.g.

$$CH_2\text{=}CH_2 + HBr \longrightarrow CH_2CH_2Br$$

bromoethane
(ethyl bromide)

They add as H^+X^-, in two stages, similarly to the addition of bromine. This time, the intermediate ion is not bridged, but is a simple carbonium ion.

$$CH_2\text{=}CH_2 + \overset{\delta+}{H}\text{—}\overset{\delta-}{Br} \longrightarrow Br^- + \underset{H}{CH_2}\text{—}\overset{+}{C}H_2 \xrightarrow{Br^-} CH_3\text{—}CH_2Br$$

An interesting problem arises with an unsymmetrical alkene such as propene because two isomeric products are possible, depending on the '*orientation*' of the addition (i.e. which way round the H and X of HX add). When addition proceeds by this heterolytic mechanism, the product obtained is the one in which the halogen becomes attached to the carbon atom with the fewer hydrogen atoms ('the less hydrogenated carbon atom'). This is known as the

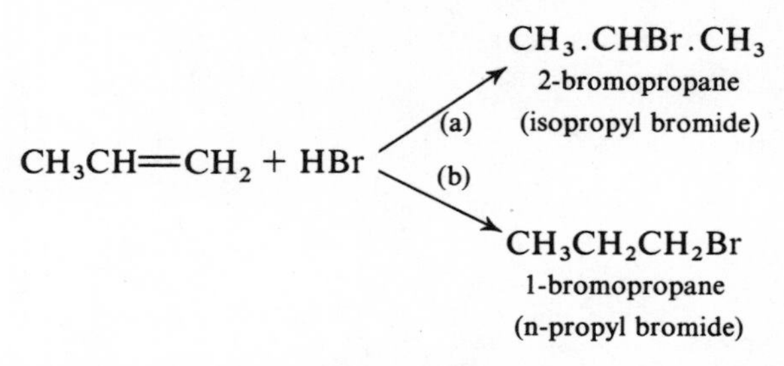

Markownikow rule. It means that with propene path (a) is followed. The reason is that in propene, for example, the methyl group repels electrons by its inductive ($+I$) effect. The π electrons are therefore encouraged to centre more on carbon-1, and the double bond becomes polarized. The electrophilic reagent, HX, tends to attack the negative end of the double bond as follows.

$$\underset{\overset{|}{\underset{X\,\delta-}{H\,\delta+}}}{CH_3{-}\overset{\delta+}{C}H{=}\overset{\delta-}{C}H_2} \longrightarrow CH_3{-}\overset{+}{C}H{-}CH_3 + X^- \longrightarrow CH_3{-}\underset{X}{\underset{|}{C}H}{-}CH_3$$

There are exceptions to this rule, however, and careful experiments have revealed that these additions can follow path (b) ('anti-Markownikow addition') under conditions which favour the formation of free radicals. These conditions are

(1) the presence of light;
(2) the presence of small quantities of added substances such as benzoyl peroxide*, which decomposes thermally to give free radicals.

$$C_6H_5CO.O.O.CO.C_6H_5 \longrightarrow 2C_6H_5CO.O\cdot$$

(3) the use of old samples of alkenes which contain peroxides formed by autoxidation. Free radicals could then be formed during reaction at higher temperatures by homolysis of these peroxides.

$$R{-}O{-}O{-}H \longrightarrow RO\cdot + HO\cdot$$

(4) the presence of oxygen which could autoxidize the alkene, to give peroxides.

To be sure of obtaining the Markownikow product, the opposites of these conditions are necessary, i.e. the exclusion of light, oxygen, and peroxide impurities, and the use of new samples of the alkenes.

It is fairly clear, then, that anti-Markownikow addition involves a free radical chain mechanism initiated by radicals formed in one of the above ways. It proceeds as follows, for hydrogen bromide addition to propene.

*The systematic name of this compound, di(benzenecarbonyl)peroxide, (cf. Chapter 11, page 116) is rarely used.

Initiation

$$\text{Light or peroxides} \longrightarrow \text{X}\cdot$$
$$\text{X}\cdot + \text{HBr} \longrightarrow \text{HX} + \text{Br}\cdot$$

Propagation

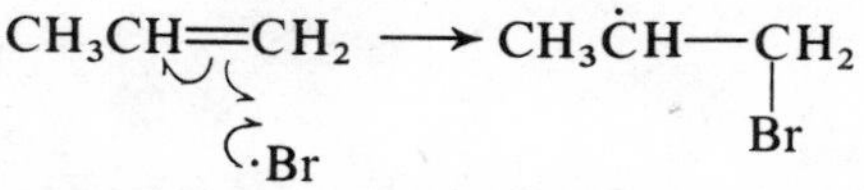

$$\text{CH}_3\text{CH}{=}\text{CH}_2 + \cdot\text{Br} \longrightarrow \text{CH}_3\dot{\text{C}}\text{H}{-}\text{CH}_2\text{Br}$$

$$\text{CH}_3\dot{\text{C}}\text{HCH}_2\text{Br} + \text{HBr} \longrightarrow \text{CH}_3\text{CH}_2\text{CH}_2\text{Br} + \text{Br}\cdot \quad \text{etc.}$$

Chains are terminated by various reactions which need not concern us, since the chains are long and hence the yields of termination products very low. (The significance of the 'half headed' arrows used in the above scheme was explained in Chapter 4, page 34.)

Because of the requirement that both stages of the propagation process should be exothermic, neither hydrogen chloride nor hydrogen iodide adds homolytically to alkenes at all readily, because with hydrogen chloride the second, and with hydrogen iodide the first, of the two stages is endothermic.

It is possible to add many reagents to alkenes by analogous homolytic mechanisms, and this is a very versatile and useful synthetic process.

Polymerization

Molecules of alkenes can add to one another by free-radical chain mechanisms once the chains have been started by an initiator like benzoyl peroxide. If the initiating radical is X·, the polymerization of an alkene $RCH{=}CH_2$ can be represented as follows.

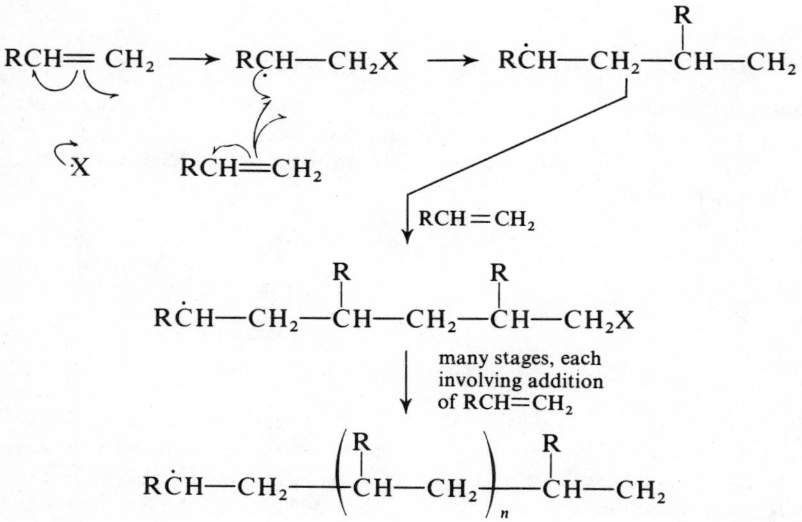

$$\text{RCH}{=}\text{CH}_2 + \cdot\text{X} \longrightarrow \text{R}\dot{\text{C}}\text{H}{-}\text{CH}_2\text{X} \xrightarrow{\text{RCH}{=}\text{CH}_2} \text{R}\dot{\text{C}}\text{H}{-}\text{CH}_2{-}\text{CH(R)}{-}\text{CH}_2\text{X}$$

$$\xrightarrow{\text{RCH}{=}\text{CH}_2} \text{R}\dot{\text{C}}\text{H}{-}\text{CH}_2{-}\text{CH(R)}{-}\text{CH}_2{-}\text{CH(R)}{-}\text{CH}_2\text{X}$$

many stages, each involving addition of $RCH{=}CH_2$

$$\text{R}\dot{\text{C}}\text{H}{-}\text{CH}_2{-}\left(\text{CH(R)}{-}\text{CH}_2\right)_n{-}\text{CH(R)}{-}\text{CH}_2$$

Very long chains can be built up in this way. The process is usually terminated by one or other of the following reactions involving two polymer radicals.

$$—CH_2—\dot{C}HR + \dot{C}HR—CH_2— \xrightarrow{\text{'dimerisation'}} —CH_2—CHR—CHR—CH_2—$$

$$—CH_2—\dot{C}HR + \dot{C}HR—CH_2— \xrightarrow{\text{'disproportionation'}} —CH_2—CH_2R + CHR{=}CH—$$

The familiar plastics which are products of 'vinyl polymerization' are built up in this way. The starting material is called the monomer, and some common polymers with their monomers are given in Table 6.1.

Synthetic rubbers are similar, but two monomers are generally polymerized together ('copolymerized'). The two monomer units then tend to alternate in the polymer chain. The most common of these is a *copolymer* of styrene and buta-1,3-diene.

The simple alkenes, ethene and propene, can be polymerized by free radical methods to give polyethene ('Polythene'), $\text{+}CH_2—CH_2\text{+}_n$ and polypropene

$$\text{+}\underset{\displaystyle CH_3}{\underset{|}{C}H}—CH_2\text{+}_n$$

respectively, the latter being similar to the former but with a higher melting point. Most commercial production of these materials, however, is not done as above, but by a heterolytic process employing the Ziegler–Natta catalysts, so called after their discoverers. These catalysts consist of two components, one of which is called the co-catalyst. A typical Ziegler–Natta catalyst is the complex of triethylaluminium $[(C_2H_5)_3Al]$ with titanium(IV) chloride $(TiCl_4)$.

Oxidation

Alkenes are readily oxidized by a number of reagents, e.g. aqueous acid or alkaline mangamate(VII) (permanganate). The decolourization of this reagent in the cold is a test for unsaturation. The product is a dihydroxy-compound (a *diol* or *glycol*).

$$CH_2{=}CH_2 + (O) + H_2O \longrightarrow \underset{\displaystyle OH}{\underset{|}{C}H_2}—\underset{\displaystyle OH}{\underset{|}{C}H_2}$$

ethane-1,2-diol
(ethylene glycol)

Alkenes are oxidized by ozone to give compounds called ozonides. Ozonized oxygen is passed into a solution of the unsaturated compound in an 'inert' solvent such as tetrachloromethane and a solution of the ozonide results. The ozonide is then either hydrolyzed with water or reduced (several reagents

Table 6.1

Monomer	*Structure*	*Polymer*	*Repeating unit*
Chloroethene (vinyl chloride)	$CH_2{=}CHCl$	Poly(chloroethene) [P.V.C., polyvinyl chloride]	$-CH_2-CHCl-$
Ethenyl ethanoate (vinyl acetate)	$CH2{=}CH.O.CO.CH_3$	Poly(ethenyl ethanoate) [P.V.A., polyvinyl acetate]	$-CH-CH(O.CO.CH_3)-$
Phenylethene (styrene)	$C_6H_5CH{=}CH_2$	Poly(phenylethene) [polystyrene]	$-CH(C_6H_5)-CH_2-$
Methyl 2-methylpropenoate (methyl methacrylate)	$CH_2{=}C(CH_3)CO.CCH_3$	Poly(methyl 2-methylpropenoate) ['Perspex', polymethyl methacrylate]	$-CH_2-C(CH_3)(CO.OCH_3)-$
Tetrafluoroethene (tetrafluoroethylene)	$CF_2{=}CF_2$	Poly(tetrafluorethene) ['Teflon', PTFE, polytetrafluoroethylene]	$-CF_2-CF_2-$

can be used) to give in each case two products, each of which contains a carbonyl ($>C{=}O$) group.

$$>C{=}C< + O_3 \longrightarrow \text{ozonide (} >C\langle O \rangle C< \text{ with } O{-}O \text{ bridge)}$$

ozonide

reduction (2H) → $>C{=}O + O{=}C< + H_2O$

hydrolysis (H_2O) → $>C{=}O + O{=}C< + H_2O_2$

Identification of the two carbonyl compounds can give valuable structural information about the alkene. This complete process (*ozonolysis*) can thus lead to a knowledge of the position of the double bond in the chain; e.g.

$$CH_3CH{=}CHCH_3 \xrightarrow{\text{ozonolysis}} 2CH_3CH{=}O$$

but-2-ene → ethanal (acetaldehyde)

$$CH_3CH_2CH{=}CH_2 \xrightarrow{\text{ozonolysis}} CH_3CH_2CH{=}O + O{=}CH_2$$

but-1-ene → propanal (propionaldehyde) + methanal (formaldehyde)

Autoxidation of compounds containing double bonds has already been mentioned (page 50). It is a complex process because the peroxides so formed can undergo homolysis to give free radicals which can then initiate polymerization of the alkenes. The final products of this process, which can take place very slowly when unsaturated compounds are exposed to air, are often therefore hard, resinous polymeric materials. This is the process which occurs when 'drying oils' like linseed oil, which contain polymerizable unsaturated constituents, are exposed to air. These drying oils, which harden gradually, are therefore used in some paints, and linseed oil particularly is used in putty.

Alkynes

Nomenclature

The rules are the same as for alkenes, but the suffix -yne is used; e.g.

$CH{\equiv}CH$ — ethyne (acetylene)

$CH_3C{\equiv}CH$ — propyne (methylacetylene)

Preparation

Alkynes can be prepared by elimination of two molecules of hydrogen halides from dihalogeno-compounds, which are easily prepared; e.g.

$$CH_3CH{=}CH_2 \xrightarrow{Br_2} CH_3CHBr.CH_2Br \xrightarrow[(-2HBr)]{\text{alcoholic KOH}} CH_3C{\equiv}CH$$

propene — propyne

$$CH_3CH{=}O \xrightarrow{PCl_5} CH_3.CHCl_2 \xrightarrow[(-2HCl)]{\text{alcoholic KOH}} CH{\equiv}CH$$

ethanal (acetaldehyde) — ethyne

Ethyne itself can be obtained by the action of water on calcium(II) dicarbide, which is prepared from lime and carbon in an electric furnace. The process is important industrially.

$$CaO + 3C \longrightarrow CaC_2 + CO$$
$$CaC_2 + 2H_2O \longrightarrow Ca(OH)_2 + C_2H_2$$

Properties and reactions

The lower members are gases; ethyne boils at −84 °C. Being an endothermic compound it is unstable, and liquid ethyne under pressure is explosive. It is therefore stored in cylinders under pressure as a solution in acetone. The cylinders have a porous lining (e.g. asbestos) in which the solution is absorbed. This method of storage is safe, and such cylinders are used to store ethyne for 'oxy-acetylene' welding. The oxy-acetylene flame is very hot (about 3000 °C).

Alkynes are very reactive and undergo many addition reactions readily. These reactions are like those of alkenes, but now, of course, take place in two stages, e.g.

$$CH{\equiv}CH + Cl_2 \longrightarrow CHCl{=}CHCl \xrightarrow{Cl_2} CHCl_2.CHCl_2$$

1,2-dichloroethene — 1,1,2,2-tetrachloroethane

Hydrogen and hydrogen halides can be added similarly in two stages. Addition of hydrogen halides follows Markownikow's rule (page 50) so that 1,1-dihaloalkanes result.

The addition of water (hydration) is important, and is not usually undergone by alkenes. Ethyne is hydrated by passing it into hot dilute sulphuric acid containing a little mercury(II) sulphate as a catalyst.

$$\begin{matrix} CH{\equiv}CH \\ + \\ H{-}OH \end{matrix} \longrightarrow [\overset{2}{C}H_2{=}\overset{1}{C}HOH]$$

vinyl alcohol

The product, ethenol (or vinyl alcohol) is one of a group of compounds called *enols* which contain a hydroxyl group and a double bond attached to the same carbon atom. These compounds readily undergo rearrangement to the usually more stable *keto*-form. The rearrangement involves migration of the hydroxyl hydrogen (it comes off as a proton and then joins on again) to carbon atom-2 (C2), with compensating movement of the double bond. In this case ethenol does not exist at all, but rearranges immediately on formation to its keto-form, which is ethanal.

$$[CH_2{=}CH{-}\overset{\displaystyle H}{\overset{|}{O}}] \longrightarrow \overset{\displaystyle H}{\overset{|}{C}}H_2{-}CH{=}O \quad (CH_3CH{=}O)$$

ethanal
(acetaldehyde)

Many compounds can be synthesized from these products, and this affords a route from carbon itself (*via* calcium(II) dicarbide) to a range of organic compounds.

Substitution reactions

Hydrogen attached to triply bound carbon can be replaced by metals. Thus, for example, if ethyne is passed into ammoniacal silver nitrate, silver(I) dicarbide (silver acetylide) $CH{\equiv}CAg$, together with the disilver compound $AgC{\equiv}CAg$ are formed. These heavy metal dicarbides are stable when wet, but easily detonated if dried.

Sodium dicarbide (sodium acetylide) is obtained by passing ethyne into the blue solution of sodium in liquid ammonia.

$$CH{\equiv}CH \xrightarrow{Na/NH_3} CH{\equiv}CNa$$

It reacts with iodomethane to give propyne.

$$CH{\equiv}CNa + CH_3I \longrightarrow CH{\equiv}CCH_3$$

Questions

1. The structural formulae of three isomers having the molecular formula C_4H_6 are shown below.

$$\begin{array}{ccccccc} H & & H & & H & & H \\ | & & | & & | & & | \\ C & = & C & - & C & = & C \\ | & & & & & & | \\ H & & & & & & H \end{array} \qquad \begin{array}{ccccccccccc} & & & & & & H & & H & & \\ & & & & & & | & & | & & \\ H & - & C & \equiv & C & - & C & - & C & - & H \\ & & & & & & | & & | & & \\ & & & & & & H & & H & & \end{array} \qquad \begin{array}{ccccccccccc} & & H & & & & & & H & & \\ & & | & & & & & & | & & \\ H & - & C & - & C & \equiv & C & - & C & - & H \\ & & | & & & & & & | & & \\ & & H & & & & & & H & & \end{array}$$

A *B* *C*

(*a*) Name the three compounds *A*, *B*, and *C*.

(*b*) (i) Draw diagrams to indicate the shape of the carbon skeleton in each of the isomers *A*, *B*, and *C*.

(ii) What explanation may be offered in terms of electronic theory for the shape of **one** of these structures?

(*c*) (i) Write the equation for the addition of excess hydrogen bromide to compound *C*.

(ii) Suggest the conditions under which this reaction would take place.

(iii) Write the equations for the reaction of this bromo-compound with

(A) aqueous potassium hydroxide.

(B) alcoholic potassium hydroxide.

(London, 1975)

2. Describe two general methods in each case which could be used for preparation of (a) an alkane, (b) an alkene.

Compare the reactions of ethene (ethylene) and ethyne (acetylene) with (i) hydrogen, (ii) chlorine, (iii) hydrogen chloride.

Describe a simple chemical test for distinguishing between ethene and ethyne.

(JMB, 1972)

3. Using a different type of starting material in each case describe briefly two methods of introducing a C=C bond into an organic structure.
Give conditions of reaction for, and explain, each of the following conversions:

(*a*) $CH_3.CH{=}CH_2 \longrightarrow CH_3.\underset{\displaystyle I}{\underset{|}{C}}H{-}CH_3$

(*b*) $CH_2{=}CH_2 \longrightarrow \underset{\displaystyle OH}{\underset{|}{C}}H_2{-}\underset{\displaystyle Br}{\underset{|}{C}}H_2$

(*c*) $CH_3{-}CH{=}CH{-}CH_3 \longrightarrow CH_3.CHO$

How would you distinguish between $CH_2{=}CH{\cdot}CH{=}CH_2$ and $CH_3{\cdot}CH_2{\cdot}C{\equiv}CH$?

(AEB, 1975)

Chapter 7

Aromatic Hydrocarbons

Benzene and some of its simple homologues have the structures, names, and boiling points given below.

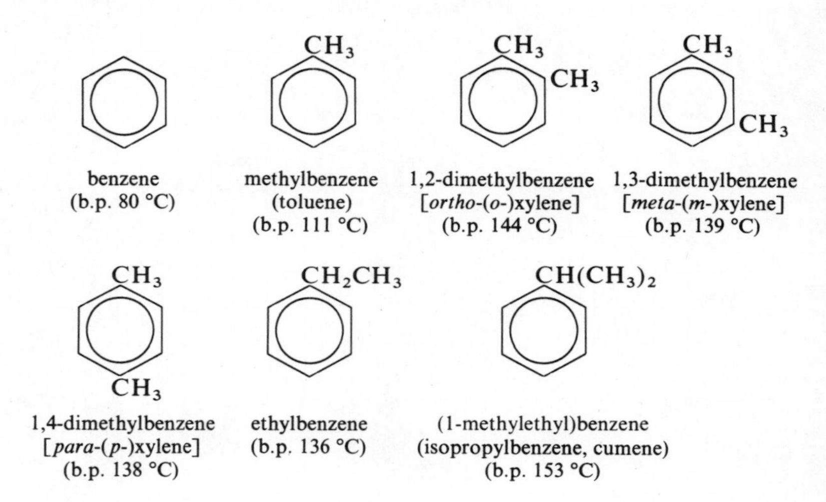

The relative positions of two substituents in disubstituted derivatives of benzene are often indicated by the prefixes *ortho-* (*o*-), *meta-* (*m*-), and *para-* (*p*-) as illustrated above for the isomeric xylenes, instead of by numbers as in the systematic names. The xylenes are, incidentally, also isomeric with ethylbenzene, and there are a number of hydrocarbons isomeric with cumene (1-methylethylbenzene) which are not illustrated. The trivial names toluene, *o*-, *m*-, and *p*-xylene, etc. are used very commonly, more frequently in fact than the systematic names, and it is essential for the student to know them. The group C_6H_5– derived by removal of hydrogen from benzene is known as the *phenyl* group, often written as Ph–. Benzene and derivatives thereof are sometimes referred to collectively as *arenes*, and the corresponding univalent groups as *aryl* groups (Ar–).

Aromatic hydrocarbons occur in coal tar, one of the products of the destructive distillation of coal. The most volatile fraction obtained by fractional distillation of coal tar contains these simple hydrocarbons. Nowadays, however, they are mainly obtained from petroleum, by catalytic cracking of the higher petroleum fractions at about 650 °C. The catalysts are metal oxides (e.g. Cr_2O_3/Al_2O_3).

Aromatic hydrocarbons are also obtained by reforming the lower petroleum fractions. This process, in which catalysts are used, is carried out under pressure at about 500 °C. The alkanes undergo cyclization and loss of hydrogen under these conditions, e.g.

$$\underset{\text{hexane}}{CH_3CH_2CH_2CH_2CH_2CH_3} \longrightarrow \underset{\text{benzene}}{C_6H_6} + 4H_2$$

$$\underset{\text{heptane}}{CH_3CH_2CH_2CH_2CH_2CH_2CH_3} \longrightarrow \underset{\substack{\text{methylbenzene}\\\text{(toluene)}}}{C_6H_5CH_3} + 4H_2$$

Further treatment of benzene homologues with hydrogen under pressure and similar catalysts converts them into benzene itself, of which larger quantities are required; e.g.

$$C_6H_5CH_3 + H_2 \longrightarrow C_6H_6 + CH_4$$

Structure and reactions of benzene

The structure of the aromatic nucleus was discussed in detail in Chapter 3, in M.O. and V.B. (resonance) terms. This unique type of structure is reflected in the properties of the benzene nucleus. Not only are these compounds considerably more stable than would be expected, as was pointed out in Chapter 3, but more particularly they do not readily undergo the addition reactions which would be expected for compounds unsaturated to the extent of three double bonds. Instead, although arenes are highly susceptible to attack by electrophiles, because of their richness in π electron density, such reactions do not generally lead to addition, but to substitution in which hydrogen is ejected and the aromatic structure preserved.

AB addition → (addition product with A and B)

AB substitution → (substituted product with A) $+ H^+ + B^-$

Addition must lead to the loss of aromatic structure because only two double bonds would be present in the product. The completeness of the conjugated system extending over the whole ring would then have been destroyed, and the extra stability resulting from the resonance of the two Kekulé forms (or in M.O. terms, from the existence of a molecular π orbital covering the whole molecule) would not be available. Thus a great thermodynamic advantage accrues from the preservation of the aromatic structure, and substitution is the preferred process.

This idea is reinforced by a comparison of the properties of benzene with cyclohexene and the cyclohexadienes, which do not possess this fully conjugated system.

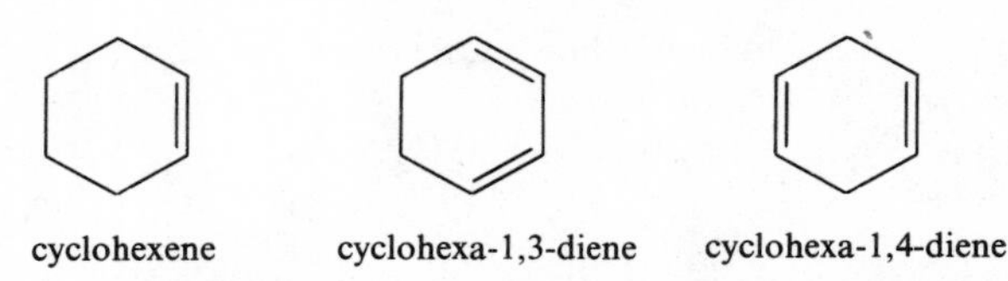

cyclohexene   cyclohexa-1,3-diene   cyclohexa-1,4-diene

They behave like alkenes and alkadienes respectively in readily undergoing addition reactions, and do not show the characteristic properties of aromatic compounds.

The delocalization of the double bonds in benzene, which is a feature of both the resonance and the M.O. pictures, is demonstrable experimentally. The important observation is that there is always only one given 1,2-(*ortho-*) disubstitution product of benzene, whereas if the double bonds were localized as in a single Kekulé structure there would be two, which would be different:

A A   A A

The condition for there being only one is that all the C–C bonds in benzene must be identical, as they are according to the views we have developed.

Addition reactions

Although the exception (substitution being the rule), a few addition reactions do occur under forcing conditions. Thus, for example, the double bonds can be catalytically hydrogenated one by one under rather vigorous conditions of temperature and pressure to give, ultimately, cyclohexane.

$\xrightarrow[\text{catalyst}]{H_2}$ $\xrightarrow[\text{catalyst}]{H_2}$ $\xrightarrow[\text{catalyst}]{H_2}$

cyclohexane

Chlorine will also add homolytically to benzene under irradiation with sunlight or ultraviolet light to give successive products.

$\xrightarrow[h\nu]{Cl_2}$ $\xrightarrow[h\nu]{Cl_2}$ $\xrightarrow[h\nu]{Cl_2}$

1,2,3,4,5,6-hexa-chlorocyclohexane ('benzene hexachloride')

The final product can exist in several stereoisomeric forms, one of which is a powerful insecticide (BHC, or 'Gammexane').

Substitution reactions

The side-chains of benzene homologues undergo substitution reactions rather like those of alkanes, because the C–H bonds concerned are of the sp^3 type. Thus, for example, as discussed in Chapter 4, methylbenzene (toluene) can be photolytically chlorinated by a homolytic mechanism analogous to that of the chlorination of methane.

We are mainly concerned, however, with reactions of the nucleus, and here, because of the high π electron density, electrophilic substitution is the most common, although there are also some important homolytic substitutions.

Electrophilic substitution

Nitration

On being warmed with a mixture of concentrated nitric and sulphuric acids, benzene is converted into nitrobenzene, a liquid (b.p. 211 °C) which smells of bitter almonds.

$$C_6H_6 \xrightarrow[60\ ^\circ C]{HNO_3/H_2SO_4} C_6H_5NO_2 + H_2O$$

$$(C_6H_6 \xrightarrow[60\ ^\circ C]{HNO_3/H_2SO_4} C_6H_5NO_2 + H_2O)$$

The mechanism of this reaction is of very great interest and has been investigated very fully. It is therefore selected as one of the reactions to be considered in more detail, and the evidence on which the accepted mechanism is based will now be summarized.

Nature of the attacking agent

The first problem was to determine the nature of the actual attacking electrophilic reagent. This turns out to be the *nitronium* cation, NO_2^+, which is formed in the mixture of concentrated acids by the following reactions.

$$HNO_3 + H_2SO_4 \rightleftharpoons H_2NO_3^+ + HSO_4^-$$

$$H_2NO_3^+ \longrightarrow NO_2^+ + H_2O$$

$$H_2O + H_2SO_4 \longrightarrow H_3O^+ + HSO_4^-$$

Adding these equations gives the complete process stoichiometrically as

$$HNO_3 + 2H_2SO_4 \longrightarrow NO_2^+ + H_3O^+ + 2HSO_4^-$$

This is supported by the following evidence.

Cryoscopic evidence

Four ions (NO_2^+, H_3O^+, $2HSO_4^-$) are formed from each molecule of nitric acid, and there should therefore be a fourfold depression of the freezing point of the solvent sulphuric acid by the solute nitric acid. This was observed. It is worth noting that in the first stage of this sequence, a proton is donated to nitric acid by sulphuric acid. In accepting the proton, nitric acid is acting as a base in sulphuric acid!

Spectroscopic evidence

The structure of the nitronium ion is:

$$O{=}\overset{+}{N}{=}O$$
$$\longleftrightarrow \quad \longleftrightarrow$$

It is expected to be a linear molecule, and the vibration which would be expected to absorb energy in an observable region of the electromagnetic spectrum is a concerted 'in-and-out' motion of the oxygen atoms relative to the nitrogen as shown. This vibration does not involve any variation in the dipole moment of the molecule, and therefore does not lead to any absorption in the infrared absorption spectrum. Such molecular vibrations can, however, be observed in the spectrum of light scattered by solutions of substances whose molecules can vibrate in this way. It is therefore in this so-called 'Raman' spectrum that one must look for an absorption due to the NO_2^+ ion. It is, in fact, observable as a single line in the Raman spectrum at a characteristic frequency.

Preparative evidence

It was found possible to prepare salts of the nitronium ion. The first to be prepared, and characterized crystallographically, was nitronium perchlorate, $NO_2^+ClO_4^-$.

Kinetic evidence

Kinetics can give information about the formation of NO_2^+ only if this process, rather than the attack of NO_2^+ on the aromatic nucleus, is rate-determining. Under most conditions, including nitration with 'mixed acid' (HNO_3/H_2SO_4), this is not the case. Instead, the formation of NO_2^+ is faster than its attack on the arene. However, if a different and less active reagent, namely nitric acid in acetic (ethanoic) acid as solvent, is used with particularly reactive aromatic substances (e.g. toluene), the formation of NO_2^+ becomes rate determining. If the above mechanism is correct, the rate should then depend only on the concentration of nitric acid, and not on that of the arene. If the reaction is then conducted with a very large excess of nitric acid, i.e.

$$[HNO_3] \gg [ArH],$$

the effect of varying nitric acid concentration is not observed during the reaction, and the kinetic order becomes zero (cf. Chapter 2, page 15). Then

$$-\frac{d[ArH]}{dt} = k_0,$$

and if [ArH], or an observed physical property which is proportional to [ArH], is plotted against t a straight line with a sharp cut-off is obtained (Figure 7.1).

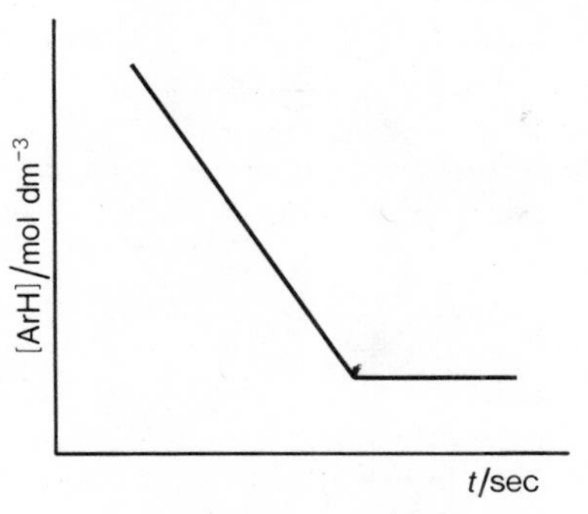

Fig. 7.1

The realization of zeroth-order kinetics shows that the reaction involves an intermediate directly derived from nitric acid by a slow process, i.e. a process involving the cleavage of a covalent bond. With these conditions, this intermediate cannot be other than NO_2^+. In the absence of sulphuric acid, it is formed as follows:

$$2HNO_3 \rightleftharpoons H_2NO_3^+ + NO_3^-$$
$$H_2NO_3^+ \longrightarrow H_2O + NO_2^+$$
$$H_2O + HNO_3 \longrightarrow H_3O^+ + NO_3^-$$

The first equilibrium, the protonation equilibrium, lies much further to the left than the corresponding reaction in the presence of sulphuric acid. The concentration of NO_2^+ is therefore much lower, and indeed, with reactive arenes when zeroth-order kinetics are observed, it can never build up because it is 'mopped up' by the arene as soon as it is formed. Thus nitric acid alone or in a solvent is a much less active nitrating agent than the nitric-sulphuric acid mixture.

Mechanism of attack

The second problem concerns the attack of NO_2^+ on the arene. Two processes must occur, namely the formation of the new bond between the NO_2 group and the nuclear carbon atom, and the cleavage of the bond between the nuclear carbon atom and the hydrogen, which must leave as a proton H^+ to balance the charges. There are two ways in which these processes can occur:

(1) The formation of the new bond and the cleavage of the old one could

occur smoothly and 'synchronously' ('during the same time'), so that the hydrogen receded as the NO_2^+ group approached, thus:

NO_2^+ H → [$\delta+$ NO_2 $\delta+$ H] → NO_2 H^+

The energy profile of the reaction would then show a maximum corresponding to this transition state (Figure 7.2).

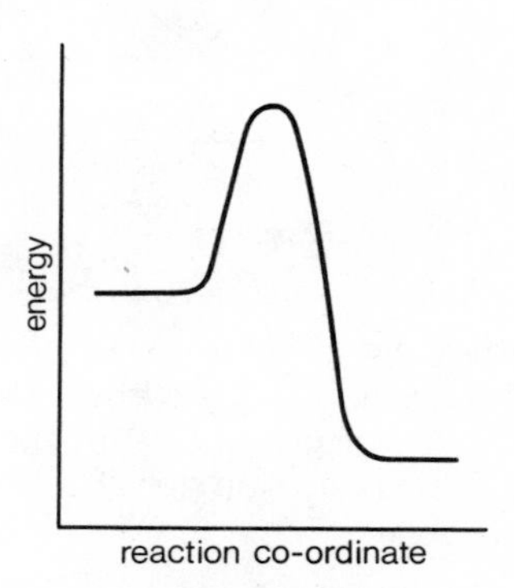

Figure 7.2

(2) Alternatively, the NO_2^+ may add to the aromatic nucleus just as a cation adds to one end of a double bond in an alkene, to give an intermediate addition complex [the so-called sigma (σ) complex]. Now, however, instead of the process being consummated by addition of an anion to the other end of the double bond, a proton is lost in a second, faster, stage to leave the nitro-compound:

$$ArH + NO_2^+ \xrightarrow{\text{slow}} ArHNO_2^+ \xrightarrow{\text{fast}} ArNO_2 + H^+$$

Within this sequence of reactions the first stage would be rate-determining. The σ-complex can be formulated as a resonance hybrid, the positive charge being delocalized among the *ortho-* (2-) and *para-* (4-) positions, thus:

O_2N H (+) ⟷ O_2N H (+) ⟷ O_2N H (+)

To indicate this, it is often written as:

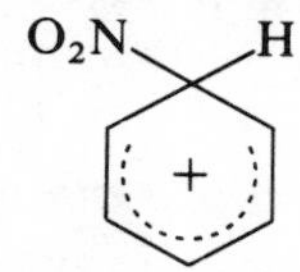

The carbon atom which is the site of attack must become sp^3 hybridized in the σ complex, whose geometry is therefore as indicated below:

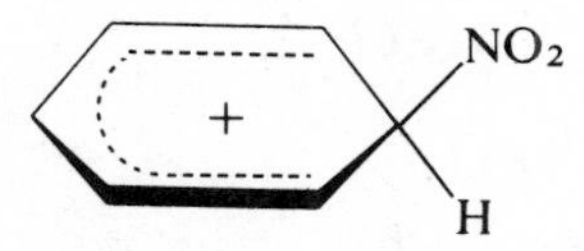

Thus the nitronium ion approaches the arene from above or below the plane of the ring, as is reasonable considering that it would be attracted by the clouds of π electron density above and below the ring, and the new bond would begin to be formed by interaction of a vacant orbital on the nitrogen with the molecular π orbital of the arene.

The resonance would be expected to confer some stability on the σ complex, which would be represented by a depression or 'trough' in the energy profile (Figure 7.3).

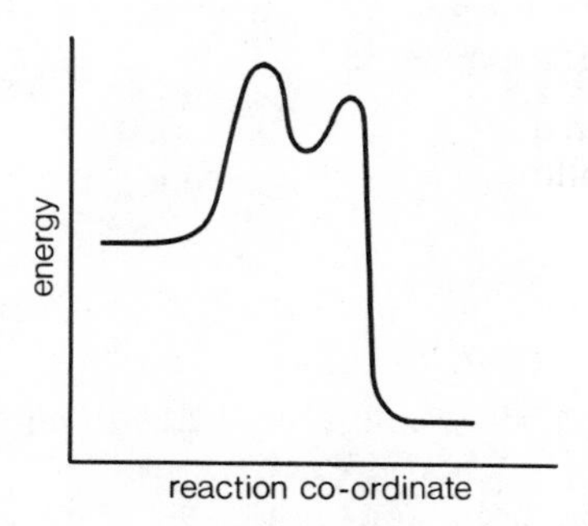

Figure 7.3

A decision between these alternatives can be made experimentally. Since the first stage of (2) is rate determining, the removal of the hydrogen plays no part in determining the rate. In (1) on the other hand the removal of the proton is part of the single stage, which must therefore be rate determining. If a hydrogen isotope, deuterium or tritium, is put into the arene in place of some of the protium atoms, its removal will be more difficult (slower) than that of protium because deuterium and tritium are heavier atoms than protium. When the experiment was done it was found that replacement of H by D or T made no difference to the overall rate of reaction, so removal of H^+ (or D^+ or T^+) did not play any part in determining the rate, and so mechanism (2) is the correct one.

This conclusion is a general one for other aromatic substitutions as well as for nitration. These reactions also proceed by these two-stage mechanisms, involving the formation of intermediate σ complexes.

Influence of substituents on the nucleus

A second nitro-group can be introduced into benzene, but with more difficulty. The product is 1,3-dinitrobenzene(*m*-dinitrobenzene).

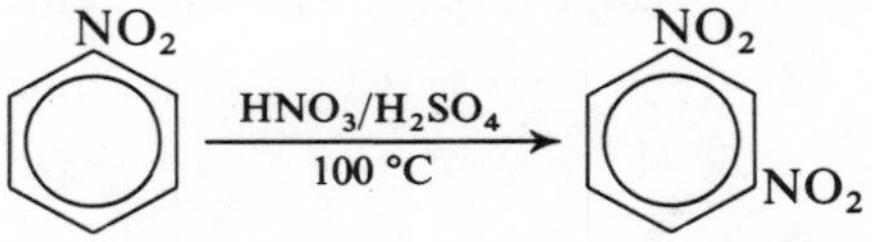

$$C_6H_5NO_2 \xrightarrow[100\,°C]{HNO_3/H_2SO_4} 1,3\text{-}C_6H_4(NO_2)_2$$

Methylbenzene (toluene) gives first a mixture of the products of nitration in the 2-(*o*-) and 4-(*p*-) positions.

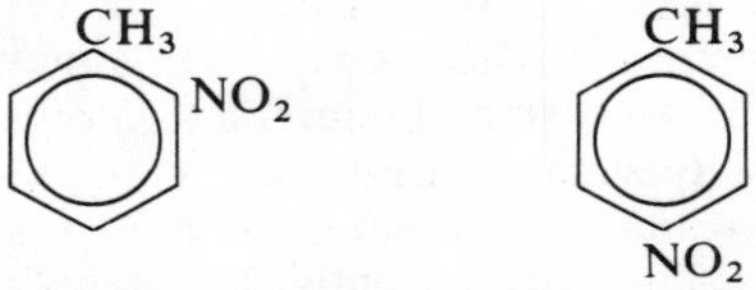

2-nitromethylbenzene
(*o*-nitrotoluene)

4-nitromethylbenzene
(*p*-nitrotoluene)

The final product of nitration of toluene is 2,4,6-trinitromethylbenzene (2,4,6-trinitrotoluene), the explosive TNT.

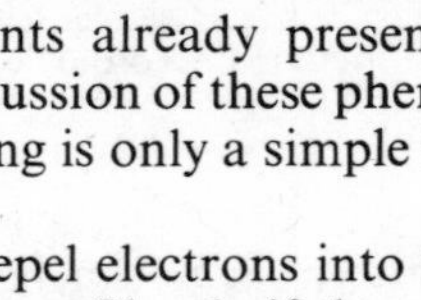

TNT

The positions at which these additional substituents enter the nucleus are determined by the substituents already present, which exert 'directing influences'. Full theoretical discussion of these phenomena involves considerable complexities, and the following is only a simple introduction to the principles involved.

Some substituent groups repel electrons into the aromatic ring, and others attract electrons out of the ring. Clearly if the net effect, as in the case of the nitro-group for example, is that of attraction, the reactivity of the nucleus as compared with unsubstituted benzene is reduced towards electrophiles such as NO_2^+, and the compound is *deactivated* towards electrophilic substitution. Thus we have already noted that a second nitro-group is more difficult to introduce than the first, because the first nitro-group deactivates the nucleus. Conversely the hydroxyl group, for example, is electron-repelling (see below), and so *activates* the nucleus towards electrophiles.

The three positions (2-, 3-, and 4-) at which further substitution can take place in a monosubstituted benzene derivative C_6H_5X are, however, not affected equally by the electron attraction or repulsion of X, and so X exerts a *directing* as well as an activating or deactivating influence. If X is OH for example, it repels electrons by virtue of the fact that the oxygen atom has two pairs of unshared electrons, which can be delocalized into the nucleus (cf. Chapter 9, page 96). These electrons can be repelled to the *ortho-* and *para-* (2- and 4-) positions as shown below.

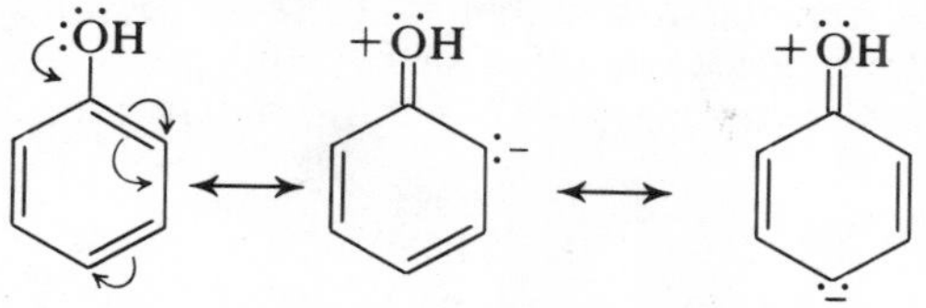

These are the positions which therefore become richer in electron density, and are the positions at which an incoming electrophile can most readily react. It is not possible to write an analogous canonical structure placing the electron pair at the *meta-* (3-) position, so that position is less reactive towards electrophiles. Nitration of phenol, which is very rapid, therefore gives almost entirely a mixture of the following two products:

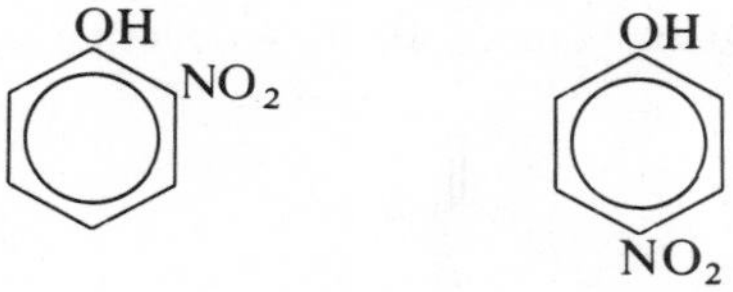

and the hydroxyl group is said to be *ortho-para* directing towards electrophiles.

The opposite effect is shown, for example, by the cyano-group, –C≡N, because it has a multiple bond between the first and second atoms. In such a group the electrons tend towards localization on the nitrogen (in this instance), and if the canonical structures corresponding to this are written it can be seen that electron pairs can move away from the *ortho-* and *para-*, but not the *meta-* positions:

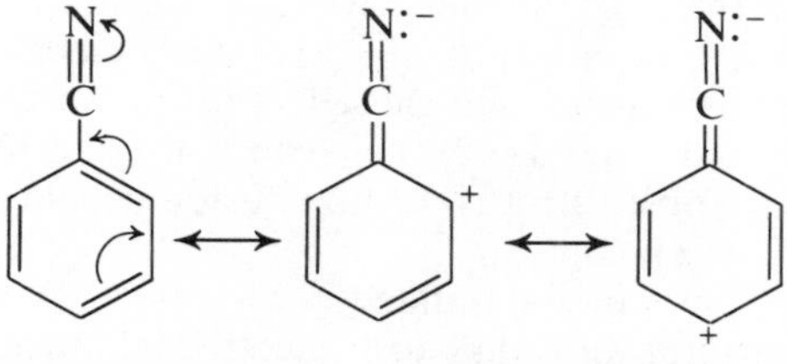

The reactivity of the *ortho-* and *para-* positions towards electrophiles is therefore reduced selectively as compared with the *meta-* position. The nucleus as a whole is deactivated but the effect is least for the *meta-* position so that the cyano-group is *meta*-directing towards electrophiles.

This mechanism of electron repulsion utilizing lone pairs on the substituent, and of electron attraction utilizing multiple bonds, is called the *conjugative* mechanism, and the groups (like OH and CN) are said to have *conjugative effects*.

There are some other groups which attract or repel electrons by a different mechanism, namely the *inductive* mechanism, which was discussed in Chapter 1 (page 9). Alkyl groups, for example, repel electrons by this mechanism, and hence activate aromatic nuclei towards electrophilic substitution. However, once a polarization such as that produced by the inductive effect of an alkyl group reaches the conjugated π electron system of an aromatic nucleus, the mechanism of its transmission within that system becomes at least partly conjugative, so that once again the *ortho*- and *para*- positions are most affected. Thus, as stated above (page 66), nitration gives the 2- and 4- isomers.

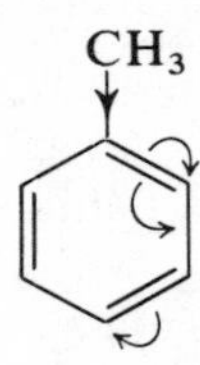

The nitro-group itself attracts electrons by both mechanisms: by the conjugative mechanism because it is unsaturated, and by the inductive mechanism because it is a dipole attached to the nucleus by its positive end (cf. Chapter 12, page 000). Therefore it is deactivating and *meta*-directing on both counts.

For all the groups so far discussed, activation goes with *ortho-para* direction and deactivation with *meta*-direction; this is the usual rule. An exceptional situation, however, occurs with the halogens which, because they are electronegative, *attract* electrons by the inductive mechanism (cf. Chapter 1, page 9) and at the same time, because they have unshared electrons, *repel* electrons by the conjugative mechanism. The net effect of this turns out to be that the nuclei of chlorobenzene and bromobenzene, for example, are deactivated towards electrophilic substitution, but substitution occurs preferentially in the *ortho*- and *para*- positions.

Although they have been discussed with reference to nitration these considerations are general, and apply to all electrophilic aromatic substitutions.

Sulphonation

If benzene is boiled under reflux with concentrated sulphuric acid, benzenesulphonic acid is formed by replacement of –H by $-SO_2.OH$, the sulphonic acid group. The sulphonation reaction is reversible.

$$C_6H_6 + H_2SO_4 \rightleftharpoons C_6H_5SO_2.OH + H_2O$$

Benzenesulphonic acid is a water-soluble deliquescent solid, m.p. 44 °C. Sulphonic acids behave as strong monobasic acids.

The mechanism of sulphonation is not so fully understood as that of nitration, but the attacking reagent is probably usually SO_3. The mechanism can be formulated as follows:

$$2H_2SO_4 \rightleftharpoons SO_3 + H_3O^+ + HSO_4^-$$

$$SO_3 + C_6H_6 \rightleftharpoons [C_6H_6(H)(SO_3^-)]^+ \rightleftharpoons C_6H_5SO_3H$$

The precise details of the last stage (the conversion of the σ complex into the product) appear to depend on the exact conditions.

The sulphonation of less reactive arenes than benzene requires the use of fuming, in place of concentrated, sulphuric acid.

Halogenation

Arenes can be halogenated by chlorine or bromine at reasonable temperatures in the presence of catalysts or 'halogen carriers'. Substances such as $FeCl_3$ or $AlCl_3$ are common halogen carriers, and iron filings are convenient because iron can be converted into the Fe(III) halide by the halogen. Thus, for example:

$$C_6H_6 + Cl_2 \xrightarrow{Fe} C_6H_5Cl + HCl$$

The function of the halogen carrier is to create a more powerful electrophile by its affinity for halide ions, e.g.

$$Cl^- + FeCl_3 \rightleftharpoons FeCl_4^-$$

Substances like $FeCl_3$, which can accept a pair of electrons and form co-ordinate bonds in this way, are called 'Lewis acids'. It is not certain whether this complex anion is fully formed, leaving the halogen cations as the active electrophiles, e.g.

$$Cl_2 + FeCl_3 \rightleftharpoons Cl^+ + FeCl_4^-,$$

or whether these exist in ion pairs, $[Cl^+ \; FeCl_4^-]$, or whether the halogen–halogen bond is extensively polarized, so that the electrophile is a complex entity such as:

$$\overset{\delta+}{Cl}\cdots\cdots\overbrace{Cl\cdots\cdots FeCl_3}^{\delta-}$$

Formulating it as if Cl^+ were the active reagent, the mechanism of the actual chlorination is as follows:

$$Cl^+ \quad C_6H_6 \longrightarrow [C_6H_6Cl]^+ \longrightarrow C_6H_5Cl \quad H^+$$

Iodine can act as a halogen-carrier in bromination by virtue of its tendency to form entities such as $Br^+\ BrI_2^-$.

Direct iodination of benzene is reversible, and this gives rise to difficulties. Other methods are therefore more usually used to prepare iodo-arenes.

Alkylation and acylation

Arenes are alkylated on treatment with alkyl halides in the presence of Lewis acids, e.g.

$$C_6H_6 + RCl \xrightarrow{AlCl_3} C_6H_5R + HCl$$

This is known as the *Friedel-Crafts* reaction, and the Lewis acid catalysts are called Friedel–Crafts catalysts. The function of the catalysts is similar to their function in halogenation, i.e. to polarize the alkyl halide and hence to generate a more powerfully electrophilic reagent, for example with the RX complex:

$$\overset{\delta+}{R}\cdots\cdots\overbrace{X\cdots\cdots AlCl_3}^{\delta-}$$

This attacks the nucleus to give a σ complex in the now familiar way.

The Friedel–Crafts reaction used in this way suffers from some serious drawbacks. It is difficult to stop when one alkyl group has entered the ring: there is a marked tendency for di- and tri-alkylbenzenes to be formed. Also the alkyl groups tend to rearrange. Thus 1-chloropropane with benzene gives 1-methylethylbenzene (cumene), and not propylbenzene as might be expected.

$$C_6H_6 + CH_3CH_2CH_2Cl \xrightarrow{AlCl_3} C_6H_5CH(CH_3)_2 + HCl$$

For these reasons it is often better to use an acyl chloride, such as ethanoyl chloride (acetyl chloride), $CH_3CO \cdot Cl$, in place of the alkyl halide. The Friedel–Crafts reaction then gives *ketones* by an analogous mechanism, e.g.

$$C_6H_6 + CH_3CO.Cl \xrightarrow{AlCl_3} C_6H_5CO.CH_3 + HCl$$

Not only does this *acylation* provide a method of synthesizing these ketones, but they can also be reduced to the corresponding hydrocarbons.

$$C_6H_5CO.CH_3 \xrightarrow{(H)} \underset{\text{ethylbenzene}}{C_6H_5CH_2CH_3}$$

Several reagents can be used to accomplish this reduction, e.g. hydrazine followed by a strong base (Wolff-Kishner), or amalgamated zinc and concentrated hydrochloric acid (Clemmensen).

Homolytic substitution

Arylation

Phenyl free radicals $C_6H_5\cdot$ can be formed by a number of methods, for example by decomposition of benzoyl peroxide at about 80 °C.

$$PhCO.O.O.CO.Ph \longrightarrow 2PhCO.O\cdot$$
$$PhCO.O\cdot \longrightarrow Ph\cdot + 2CO_2$$

If this is done in solution in an arene (e.g. benzene), these radicals attack the arene to give a σ complex in which the unpaired electron is delocalized like the positive charge in electrophilic substitution. Removal of a hydrogen *atom* (not a proton) from this gives the substitution product, which is a biaryl, in this case biphenyl $C_6H_5C_6H_5$.

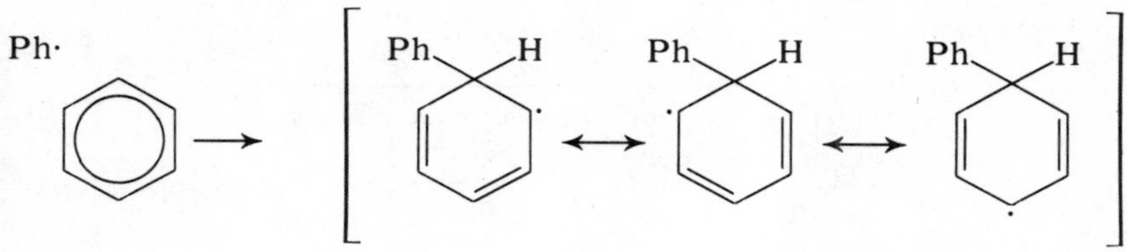

Ph· Ph H Ph H Ph H

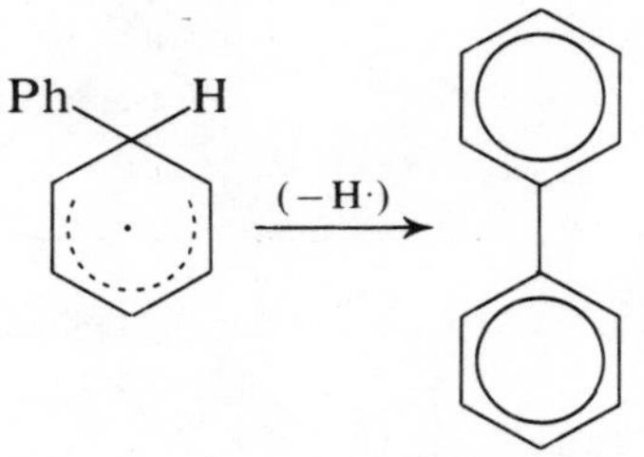

Ph H (−H·)

biphenyl

Some dehydrogenating agent is necessary to remove the hydrogen atom from the σ complex, and this part of the process, although now fairly well understood, can be complicated. The stoichiometric result is that some of this hydrogen can be accounted for as benzoic acid, PhCO.OH, which is also formed, but the mechanism of even this process is neither simple nor immediately apparent.

The reaction is quite general. Aryl radicals other than phenyl can be used, and also arenes other than benzene.

Questions

1. (*a*) What is meant by an *electrophile*?
   (*b*) In the nitration of benzene to form mononitrobenzene,
      (i) State the reagents and essential reaction conditions.
      (ii) Write the name and formula of the electrophile in this reaction.
      (iii) Write the equation for the reaction by which the electrophile is formed.
      (iv) Give **one** piece of evidence for the existence of the above electrophile.
   (*c*) **Nitrobenzene $\xrightarrow{I}$ Aniline (phenylamine) $\xrightarrow{II}$ Iodobenzene.**
   State **briefly** the reaction conditions and the reagents required to carry out stages I and II in the above reaction scheme. (**Note**—*details of the isolation and purification of the products are* **not** *required.*)

(WEJC, 1975)

2. Summarize the reactions, if any, of simple alkanes, alkenes, alkynes and of benzene with the following reagents:
   (*a*) bromine,
   (*b*) potassium permanganate, and
   (*c*) sulphuric acid.
   Clear structural formulae of reaction products should be given and approximate experimental conditions stated.
   For each of the types of hydrocarbon, select one reaction and briefly discuss its mechanism.

(London, 1975)

3. In the Journal of the Chemical Society for 1950, Hughes, Ingold and Reed report some kinetic studies on aromatic nitration. In one experiment acetic acid containing 0.2 per cent water was used as solvent and pure nitric acid was added to make a 7 M solution.

   The kinetics of the nitration of ethylbenzene, C_6H_5—C_2H_5, were studied with the following results at 20 °C:

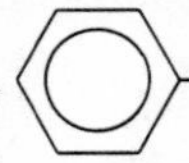

Time/min	*Concentration of ethylbenzene /mol dm^{-3}*
0	0.090
8.0	0.063
11.0	0.053
13.0	0.049
16.0	0.037
21.0	0.024
25.0	0.009

   Determine an order of reaction from these results. What does this suggest about the mechanism of the reaction?
   What products are likely?
   Briefly review the various practical techniques used in the study of rates of reaction and comment on their suitability for the study of this reaction.

(Nuffield, 1975)

Chapter 8

The Halogen Function

Nomenclature

Halides in general are named as halogeno-derivatives of hydrocarbons, e.g.,

CH_3CH_2Cl
chloroethane
(ethyl chloride)

$(CH_3)_2CHBr$
2-bromopropane
(isopropyl bromide)

$CH_3CH_2CH_2Br$
1-bromopropane
(n-propyl bromide)

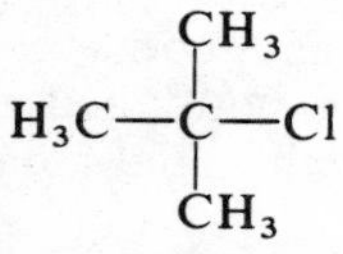

$H_3C-C(CH_3)_2-Cl$
2-chloro-2-methylpropane
(t-butyl chloride)

$CH_3CH_2CHCl.CH_3$
2-chlorobutane
(sec-butyl chloride)

$CH_3CH_2CH_2CH_2Cl$
1-chlorobutane
(n-butyl chloride)

$(CH_3)_2CHCH_2Cl$
1-chloro-2-methylpropane
(isobutyl chloride)

$\overset{2}{C}H_2{=}\overset{1}{C}HCl$
chloroethene
(vinyl chloride)

$\overset{3}{C}H_2{=}\overset{2}{C}H{-}\overset{1}{C}H_2Cl$
1-chloroprop-2-ene
(allyl chloride)

C_6H_5Cl
chlorobenzene

The trivial names (in parentheses) for simpler halides are, however, used very frequently, particularly for the above examples. Alkyl groups of the type RCH_2- are called *primary*–, $RR'CH-$ *secondary* (sec-), and $RR'R''C-$ *tertiary* (t-). The prefix iso- is reserved for the particular branch $(CH_3)_2CH-$ at the end of a chain. Thus of the two isomeric propyl radicals, *n*-propyl is primary and isopropyl secondary. Of the four isomeric butyl radicals, *n*- and isobutyl are primary, sec-butyl is secondary, and t-butyl tertiary. Note, however, that these are not necessarily named systematically as halobutanes, but rather by reference to the longest unbranched chain they contain.

Dihalides in which the two halogens are attached to the same carbon are called *geminal* (gem-) dihalides and the bivalent groups containing them are called alkylidene groups, e.g.

CH_2CHCl_2
1,1-dichloroethane
(ethylidene chloride)

$CH_3CCl_2CH_3$
2,2-dichloropropane
(isopropylidene chloride)

Dihalides in which the two halogens are attached to *adjacent* carbons are called *vicinal* (vic-) dihalides, e.g. $CH_2Br \cdot CH_2Br$, 1,2-dibromoethane (ethylene dibromide).

Fluoro-compounds

Fluoro-compounds are of considerable importance both scientifically and industrially. While they share much of the chemistry of other halogeno-compounds, however, they do form a rather specialized branch of organic chemistry because fluorine is in some ways an unusual element, partly on account of its extreme electronegativity. The study of organofluorine compounds is therefore in the main beyond the scope of this book, and this chapter is concerned only with the halogens chlorine, bromine, and iodine.

Introduction of the halogen function into carbon frameworks

The reactions in which halogeno-compounds are formed are more appropriately discussed in the contexts of the starting materials (the substances which undergo these reactions). These reactions are therefore merely mentioned here for completeness and the reader is referred to the more detailed discussions.

Halogenation of hydrocarbons, i.e. replacement of hydrogen by halogen, gives halogeno-compounds. Thus halogenation of alkanes gives halogeno-alkanes (Chapter 5, page 40) and of arenes gives halogeno-arenes (Chapter 7, page 69). Alkenes can conviently be brominated specifically at the carbon atom adjacent to the double bond (the 'allylic' position) by the specific reagent *N*-bromosuccinimide (NBS) in boiling tetrachloromethane in the presence of free-radical initiators.

$CH_2.CO$
NBr
$CH_2.CO$
N-bromosuccinimide

The mechanism is of the 'radical-chain type', e.g.

cyclohexene $\xrightarrow{NBS}$ 1-bromocyclohexane (—Br)

Addition reactions also give halides, e.g. addition of bromine to alkenes gives vic-dibromides (Chapter 6, page 48) and addition of hydrogen halides gives monohalogeno-compounds (Chapter 6, page 49).

Hydroxyl groups can conveniently be replaced by halogens, by the use of particular reagents (Chapter 9, pages 92, 93).

Properties and reactions of halogeno-compounds

Halides are generally insoluble in water, and of higher boiling point than the parent hydrocarbons. Usually the boiling points increase with increasing relative atomic mass of the halogen (F < Cl < Br < I), e.g. chloroethane has b.p. 13.1 °C, iodoethane has b.p. 72.3 °C, chlorobenzene has b.p. 132 °C.

Halogen functions are rather reactive but the reactions depend on the framework to which the halogen is attached. Thus chloroalkanes and chloroarenes have some, but by no means all, reactions in common. Side chain aromatic halides, like (chloromethyl)benzene (benzyl chloride), $C_6H_5CH_2Cl$, behave much more like halogeno-alkanes, but are often rather more reactive than them. The similarity is due to the attachment of the halogen to sp^3 carbon in both types of compounds, and the enhanced reactivity arises from the presence of unsaturation at one remove, as it were, from the halogen. A similar situation arises in allyl halides and these also show this enhanced reactivity.

$C_6H_5CH_2Cl$

(chloromethyl)benzene
(benzylc chloride)

$\overset{\gamma}{C}H_2{=}\overset{\beta}{C}H{-}\overset{\alpha}{C}H_2Cl$

1-chloroprop-2-ene
(allyl chloride)

Halogeno-arenes behave differently because the halogen is attached to sp^2 carbon. They are in some respects more similar to vinyl halides (e.g. chloroethene), which also have this structural feature.

$\overset{\beta}{C}H_2{=}\overset{\alpha}{C}HCl$

chloroethene
(vinyl chloride)

The result of this is that nucleophiles in particular find it difficult to approach and form a bond with the carbon atom to which the halogen is attached (often called the α-(alpha-) carbon atom) because of the high π electron density associated with the unsaturated systems. These halides, therefore, are not generally very susceptible to reactions involving attack by nucleophiles. Thus α,β-(alpha-, beta-) unsaturation as in vinyl halides *reduces* reactivity towards nucleophiles, whereas the β,γ-beta-, gamma-) unsaturation occurring in benzyl or allyl halides *enhances* it.

The main types of reactions of the halogen function can now be classified and discussed.

Reduction

Halogens, more particularly iodine, attached to sp^3 carbon, can be replaced by hydrogen by strong reduction, e.g. with hydrogen iodide.

$$RI + HI \longrightarrow RH + I_2$$

Reactions with metals

Solutions of halogeno-compounds in ether ('ethereal solutions' of them) react with sodium. The halogen is lost and union of the alkyl groups takes place, although the yield varies greatly with the particular halide.

$$2RX + 2Na \longrightarrow R{-}R + 2NaX$$

Although the mechanism of this *Wurtz* reaction is still a matter of some discussion, it is probably a homolytic process under many conditions. If a mixture of two different haloalkanes (RX and R′X) is used, the three possible products RR, RR′, and R′R′ are formed. If RX and R′X were present in equal concentrations, ideally the yields of RR, RR′, and R′R′ respectively would be in the ratio 1:2:1 because RR and R′R′ can each be formed only in one way (by union of R andR or R′ and R′) whereas the mixed product can be formed in two ways (by union of R with R′ or of R′ with R). This prediction is not strictly borne out in practice because considerations other than purely statistical ones influence the course of the reaction.

The reaction can also be used with mixtures of alkyl and aryl halides, although the symmetrical products as well as the usually desirable alkylbenzenes are also formed, e.g.

$$C_6H_5Br + CH_3CH_2Br \xrightarrow{Na} \underset{\text{ethylbenzene}}{C_6H_5CH_2CH_3}$$

In this modification it is known as the *Wurtz–Fittig* reaction.

A similar process is the *Ullmann* reaction, which is used to link together two aryl groups. In this the aryl halide is heated with copper powder.

$$2C_6H_5I + 2Cu \longrightarrow \underset{\text{biphenyl}}{C_6H_5{-}C_6H_5} + 2CuI$$

Alkyl and aryl halides in solution in ether undergo a most important reaction with magnesium to give alkyl- and aryl-magnesium halides, the so-called *Grignard* reagents. The reaction is sometimes a little difficult to start, but once started proceeds rapidly. It is essential that the reagents are dry. Warming with the hand is sometimes sufficient to start the reaction, but often addition of a crystal of iodine is necessary. Once started the rapidity of the reaction can combine with its exothermicity and the volatility of the ether solvent to create a fire hazard. The reaction must therefore be carefully controlled by using an efficient reflux condenser and by adding the ethereal solution of the halide in small quantities.

$$\text{R—X} + \text{Mg} \xrightarrow[\text{ether}]{\text{dry}} \text{'RMgX'}$$

The Grignard reagents are seldom isolated, but are used for further reactions in the ethereal solution. These solutions must be protected from moisture and from the atmosphere because the Grignard reagents are readily oxidized and hydrolyzed.

The reactions of Grignard reagents depend on the fact that they are polarized by the electropositive metal so that they react as if they had the structure R^-MgX^+. Negative ions like R^- are called *carbanions* and are very powerful nucleophiles. The importance of Grignard reagents is as a source of carbanions. The exact constitution of them in ethereal solution has been much discussed; certainly it is complex and the solutions do not contain *free* carbanions. It is sufficient for present purposes to consider that they react as if they were R^-MgX^+. It is then easy to see that hydrolysis will give hydrocarbons.

$$\begin{matrix} \overset{\delta-}{\text{R}}\text{—}\overset{\delta+}{\text{MgX}} \\ \overset{\delta+}{\text{H}}\text{—}\overset{\delta-}{\text{OH}} \end{matrix} \longrightarrow \text{RH} + \text{Mg(OH)X}$$

Alkyl and aryl Grignard reagents are extremely versatile synthetic reagents and can be used to synthesize, for example, alcohols and carboxylic acids as well as hydrocarbons. These reactions are discussed in more detail in their appropriate contexts (cf. Chapter 10, page 89; Chapter 11, page 107).

Halogeno-compounds also react with many other metals. Lithium compounds, which can be prepared by treatment of the halide with lithium as well as by other methods, are of considerable use.

$$\text{n-}C_4H_9Cl + 2Li \xrightarrow{\text{hexane}} \underset{\text{n-butyl-lithium}}{\text{n-}C_4H_9Li} + LiCl$$

They undergo reactions analogous to those of Grignard reagents, but often more readily, reacting as $R^- Li^+$.

The reason why aryl as well as alkyl halides undergo these reactions with metals is that the metallic reagents, being electropositive, are not nucleophiles, whose approach would be inhibited by the π electron cloud.

Nucleophilic replacement of halogens

For the reason already stated, it is in general halogens attached to sp^3 carbon that are susceptible to replacement by nucleophiles in reactions such as:

$$Y^- + RX \longrightarrow RY + X^-$$

Aryl halides undergo these reactions only very reluctantly because the high π electron density around the nucleus inhibits the approach of nucleophiles, unless special structural features (strongly electron-attracting groups) are present which increase the reactivity. With simple aryl halides, like chlorobenzene, extremely vigorous conditions are required, e.g.

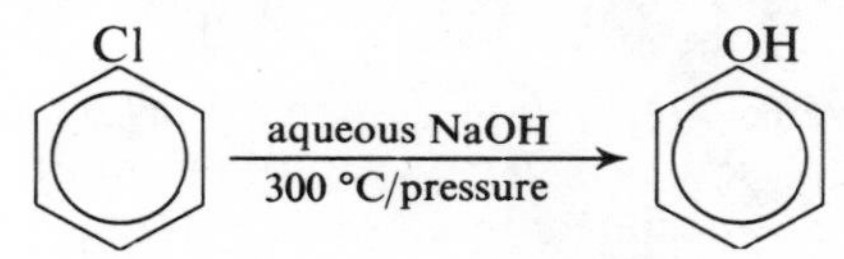

Such reactions cannot be conveniently carried out in the laboratory. Vinyl halides behave similarly, and for the same reason.

With alkyl, allylic, and side chain aromatic halides however, these reactions are much easier and take place readily under ordinary laboratory conditions.

There are very many nucleophilic reagents, many of which are anions, particularly those which are strong bases, i.e. those whose conjugate acids are weak acids.

	Nucleophile	*Conjugate acid*	
Decreasing nucleophilicity ↓	OR^- (alkoxide ions)	ROH (alcohols)	Increasing strength of acids ↓
	OH^-	H_2O	
	CN^-	HCN	
	NO_2^-	HNO_2	
	NO_3^-	HNO_3	
	Hal^-	HHal	

Thus hydroxide and most alkoxide ions are powerful nucleophiles, but chloride and bromide ions are much less powerful. This correlation of basic strength with nucleophilicity is not always obeyed, however, because there may be other factors which can detract from the nucleophilicity of a strong base. Thus, for example, the 2-methylprop-2-oxide ion (usually called the 't-butoxide ion'), $(CH_3)_3CO^-$, is a strong base because it is the anion of the very weak acid t-butyl alcohol (systematically, 2-methylpropan-2-ol), $(CH_3)_3COH$*. It is, however, very bulky and its size prevents it from reacting readily at tetravalent carbon because its access to the reaction site is impeded. It is therefore a weaker nucleophile than would be expected. However, as long as it is accepted with reservations such as this, the above correlation is a reasonable guide.

A variety of products can be obtained by nucleophilic replacement of halogens with these reagents, e.g.

$$RX + OH^- \longrightarrow \underset{\text{alcohols}}{ROH} + X^-$$

$$RX + OR'^- \longrightarrow \underset{\text{ethers (alkoxyalkanes)}}{ROR'} + X^- \text{ (the \textit{Williamson} synthesis)}$$

$$RX + CN^- \longrightarrow \underset{\text{cyanides (nitriles)}}{RCN} + X^-$$

* See Chapter 9, pages 86, 87 for systematic naming of hydroxy-compounds.

$$RX + NO_2^- \longrightarrow R{-}O{-}N{=}O \left(+ R{-}\overset{+}{N}\begin{matrix} O^- \\ \\ O \end{matrix} \right) + X^-$$

nitrates (+ nitroalkanes)

$$RX + NO_3^- \longrightarrow \underset{\text{nitrates}}{RNO_3} + X^-$$

$$RCl + I^- \longrightarrow \underset{\text{iodides}}{RI} + Cl^-$$

These reactions are sometimes catalyzed by silver ions. Thus silver nitrate is usually used to prepare alkyl nitrates, and moist silver oxide, which reacts as silver hydroxide (AgOH), is sometimes a convenient reagent for the preparation of alcohols (cf. Chapter 9, page 87). Silver nitrite gives mixtures of the alkyl nitrites and the isomeric nitroalkanes (cf. Chapter 12, page 141).

Nucleophiles are not necessarily anions. Thus water and alcohols can act as nucleophiles, and ammonia and amines (i.e. substituted ammonias in which one or more hydrogen atoms have been replaced by alkyl or aryl groups) are very important nucleophiles. The products of these reactions are quaternary ammonium cations, as their salts with the anion provided by the displaced halogen, e.g.

$$RX + NH_3 \longrightarrow R\overset{+}{N}H_3X^-$$

$$RX + NR'_3 \longrightarrow RR'_3\overset{+}{N}X^-$$

Thus the requirement for a nucleophile is not necessarily negative charge, but the presence of an unshared pair of electrons which it can donate to form the new bond. Thus, for example, ammonia is a nucleophile for the same reason as it is a base, i.e. that it can donate its unshared electron pair to form a bond by coordination, with a proton when acting as a base:

$$\ddot{N}H_3 + H^+ \rightleftharpoons NH_4^+,$$

and with carbon when acting as a nucleophile.

Mechanisms

The mechanism of these reactions ('substitution at a saturated carbon atom') has been very intensively studied, notably by Hughes and Ingold. It is well understood and is therefore selected as the second reaction for detailed study from the point of view of the establishment of its mechanism.

There are two mechanisms which can operate.

(1) $$RX \underset{}{\overset{\text{slow}}{\rightleftharpoons}} R^+ + X^- \quad \text{(rate-determining)}$$

$$R^+ + Y^- \xrightarrow{\text{fast}} RY$$

The first stage is rate-determining and since only one molecule takes part in it, the mechanism is unimolecular. Typically it gives first order kinetics:

$$-\frac{d[RX]}{dt} = k_1[RX]$$

It is known as the S_N1 mechanism (substitution, nucleophilic, unimolecular), and, as shown, involves carbonium ions R^+ as intermediates.

(2) $$\ddot{Y}^- + R—X \longrightarrow [\overset{\delta-}{Y}\cdots\cdots R\cdots\cdots \overset{\delta-}{X}] \longrightarrow YR + X^-$$

In this mechanism the approach of Y and the recession of X are synchronous, and the process occurs in one smooth stage, passing through the transition state illustrated. It is therefore bimolecular, since both Y^- and RX are involved in the single stage (which must of course be the rate determining stage, since there are no other stages). If changes in the concentrations of both reagents can be observed, this will give second order kinetics:

$$-\frac{d[RX]}{dt} = k_2[RX][Y^-]$$

However, if the reagent is present in large excess so that variations in its concentration are insignificant, the kinetics will reduce to first order as explained in Chapter 2. This can be arranged by so designing the experiment, and it is often done to simplify the kinetics. However, in a solvolytic reaction, for example a hydrolysis carried out in a partly aqueous solvent in which water is the nucleophile, this situation cannot be avoided, and first-order kinetics must result.

$$RX + H_2O \longrightarrow R\overset{+}{O}H_2 + X^-$$

$$\text{(followed by: } R\overset{+}{O}H_2 \longrightarrow ROH + H^+\text{)}$$

Thus the two mechanisms cannot be distinguished kinetically in such solvolyses, and other, non-kinetic data are required to diagnose the mechanism (cf. Chapter 11, pages 136–7).

This mechanism is known as the S_N2 (substitution, nucleophilic, bimolecular mechanism).

The mechanisms have been illustrated for negatively charged nucleophiles. For neutral nucleophiles, e.g. NR'_3, they can be written as follows, but are essentially the same mechanisms.

S_N1 $$R—X \underset{}{\overset{\text{slow}}{\rightleftharpoons}} R^+ + X^-$$

$$R^+ + R'_3N \xrightarrow{\text{fast}} R'_3\overset{+}{N}R$$

S_N2 $$R'_3\ddot{N} + R—X \longrightarrow [R'_3\overset{\delta+}{N}\cdots\cdots R\cdots\cdots\overset{\delta-}{X}] \longrightarrow R'_3\overset{+}{N}R + X^-$$

Although the actual charges are different, the way in which they change (the way in which the electron-pairs move) during reaction is the same as with anionic nucleophiles.

The course of these reactions can be influenced by a number of factors, and the two mechanisms respond differently.

1. *The nature of R.*—The S_N1 reaction involves full dissociation of RX in the rate determining stage and this is greatly assisted by electron repulsion towards X. Thus any electron-repelling substituent in R will increase the rate of S_N1 reactions. In particular, since alkyl groups themselves repel electrons, the more branched the chain the greater the repulsion. Thus electron repulsion (+I) increases from primary to tertiary alkyl groups, e.g.

$$CH_3\rightarrow CH_2\rightarrow \;<\; \begin{array}{l} CH_3\searrow \\ \quad\quad CH\twoheadrightarrow \\ CH_3\nearrow \end{array} \;<\; \begin{array}{l} CH_3\searrow \\ CH_3\rightarrow C\twoheadrightarrow\!\!\rightarrow \\ CH_3\nearrow \end{array}$$

The S_N2 reaction is much less affected by electron repulsion of this kind. It is in fact usually slightly hindered because the greater concentration of negative charge in the vicinity of the α-carbon atom tends to inhibit the approach of the nucleophile, and this is part of the rate-determining stage in this mechanism.

The result is that, under ordinary conditions, primary halides such as ethyl halides react by S_N2 mechanisms and tertiary halides by S_N1 mechanisms. Secondary halides (e.g. 2-halopropanes) can react by either mechanism, depending on the conditions. For example, the mechanism can be S_N2 at high concentrations of the nucleophile, and S_N1 at low concentrations, since only the S_N2 mechanism is affected by the nucleophile concentration. Since aryl groups are electron rich and hence electron repelling, (halomethyl)-benzenes (benzyl halides) although primary behave like *secondary* halo-alkanes, as also do allyl halides (e.g. 1-haloprop-2-enes) for a similar reason.

2. *The nature of Y.*—The nucleophile is involved in the rate-determining stage of the S_N2 mechanism, but not in that of the S_N1. Consequently, the S_N2 mechanism is favoured by the use of a more powerful nucleophile. Moreover, since the rates of S_N1 reactions are independent of the nucleophile, quite feeble nucleophiles such as water can be used to accomplish reactions of tertiary halides which react by S_N1 mechanisms. Thus, for example, the rate of hydrolysis of 2-chloro-2-methylpropane (t-butyl chloride) in a partially aqueous solvent:

$$\text{t-BuCl} + H_2O \longrightarrow \text{t-BuOH} + H^+ + Cl^-$$

is not increased by the addition of alkali, whereas the corresponding S_N2 reaction of a haloethane, for example, is highly dependent on the concentration of alkali present. (Frequently, as in the above equation, the symbol t-Bu is used to represent the 1,1-dimethylethyl (t-butyl) group,

$$\begin{array}{l} H_3C\diagdown \\ H_3C-C- \\ H_3C\diagup \end{array}).$$

3. *The nature of the solvent.*—The rates of S_N1 reactions are usually greatly increased by more strongly ionizing solvents, because the mechanism involves dissociation of RX to give ions in the rate determining stage. The S_N2 reaction is normally not nearly so sensitive to solvent. Thus it is sometimes possible (e.g. with secondary halides such as 2-halopropanes) to cause the mechanism to change from S_N2 to S_N1 by making the solvent more polar (e.g. by adding more water to it).

There are other differences between S_N1 and S_N2 mechanisms. One in particular is that their stereochemical consequences are different, and such differences can be used to diagnose the mechanism in those cases (e.g. solvolyses) when the kinetic criterion does not distinguish between them. These stereochemical aspects are dealt with in Chapter 11 (page 116).

Elimination reactions

The elimination of hydrogen halides from halogenoalkanes to give alkenes was mentioned in Chapter 6 as a method for the introduction of a double bond. The usual reagent is alcoholic potash (which contains ethoxide ions), e.g.

$$(H_3C)(CH_3)(^{\beta}H_3C)\overset{\alpha}{C}{-}Cl + OEt^- \longrightarrow (H_3C)_2C{=}CH_2 + HOEt + Cl^-$$

2-chloro-2-methylpropane (t-butyl chloride) → 2-methylpropene

The reagent must this time be essentially a base, because its function is to remove a hydrogen attached to a β-carbon atom (i.e. a β-hydrogen atom) *as a proton.* Thus, in the above example, the electron pair for the H–OEt bond is supplied by the oxygen of the EtO^- ion.

Mechanisms

There are two mechanisms for elimination reactions, which are analogous to the S_N1 and S_N2 mechanisms for substitution. The difference is that they both involve attack by a *base* on the *β-hydrogen,* instead of a *nucleophile* on the *α-carbon.* The halide is written in general as $H\overset{(\beta)}{C}R_2.\overset{(\alpha)}{C}R_2X$, where the groups R can be hydrogen or alkyl groups.

E1 (elimination, unimolecular)

$$HCR_2{-}CR_2X \underset{}{\overset{\text{slow}}{\rightleftharpoons}} HCR_2{-}\overset{+}{C}R_2 + X^- \quad \text{(rate determining)}$$

$$\ddot{B}^- + H{-}CR_2{-}\overset{+}{C}R_2 \xrightarrow{\text{fast}} BH + CR_2{=}CR_2$$

E2 (elimination, bimolecular)

$$\ddot{B}^- + H{-}CR_2{-}CR_2{-}X \rightarrow [\overset{\delta-}{B}\cdots H\cdots CR_2 \overset{\cdots}{=} CR_2\cdots \overset{\delta-}{X}] \rightarrow$$

$$BH + CR_2{=}CR_2 + X^-$$

The case illustrated is the most usual one, in which the base is an anion, B^-. The product BH, its conjugate acid, is then a neutral molecule. If B were neutral, the conjugate acid would be positively charged, BH^+.

Elimination must always be in competition with substitution because the base is also a nucleophile and can in principle also attack the α-carbon giving, in the above scheme, a substitution product $HCR_2{-}CR_2B$.

$$HCR_2CR_2X + B^- \xrightarrow{\text{elimination}} BH + CR_2{=}CR_2 + X^-$$

$$HCR_2CR_2X + B^- \xrightarrow{\text{substitution}} HCR_2CR_2B + X^-$$

Elimination is of course favoured by the use of strong bases, and that is why alcoholic potash is a good reagent for eliminations.

Elimination occurs more readily with tertiary, and to a lesser extent secondary, halides, while substitution is the preferred reaction with primary halides. Thus 2-halo-2-methylpropanes (t-butyl halides) give almost entirely the alkene 2-methylprotene with ethoxide ions, as shown above. 2-Halopropanes (isopropyl halides) give both propene and 2-ethoxypropane (isopropyl ethyl ether) (the substitution product), although the alkene is the major product.

$$(H_3C)_2CH{-}Br + OEt^- \rightarrow CH_3CH{=}CH_2 + HOEt + Br^- \quad \text{major}$$

$$(H_3C)_2CH{-}Br + OEt^- \rightarrow (H_3C)_2CHOEt + Br^- \quad \text{minor}$$

Haloethanes, however, give almost entirely ethoxyethane (diethyl ether) and virtually no ethene.

$$CH_3CH_2Br + OEt^- \longrightarrow CH_3CH_2OCH_2CH_3 + Br^-$$

Part of the reason for this may be that substitution is more difficult in the more branched halides compared with elimination, because the reaction site (the α-carbon atom) is more crowded by substituent groups, and access to it by the reagent is more difficult. The β-hydrogen is less crowded and access to it is not so restricted. Although this may be a contributory factor, it is certainly not the whole story. The reasons are understood, and a detailed analysis may be found in more advanced studies.

Questions

1. Explain the statement that the conversion of benzene to bromobenzene is an example of electrophilic substitution. Show how the reagents employed in this halogenation reaction react to produce the electrophile.
   Outline the preparation of bromobenzene from benzene in the laboratory. Diagrams are *not* required.
   The action of a halogen on an alkane to produce haloalkanes is different from that with benzene. Summarize the mechanism by which the attack of chlorine on methane proceeds.
   Give **one** reagent which undergoes a similar reaction with an aliphatic and an aromatic halide, and **two** which do not. Write equations for any reactions.
   (WJEC, 1974)

2. Describe, with full experimental details, the preparation of pure bromoethane (ethyl bromide) from ethanol (ethyl alcohol).
   How, and under what conditions, does ethyl bromide react with: (a) potassium cyanide, (b) ammonia, (c) potassium hydroxide, (d) sodium? Give the equation and name the organic product in each case.
   (JMB, 1973)

3. 1,2-Dibromoethane may conveniently be prepared on a laboratory scale by passing ethene (ethylene) gas into liquid bromine covered by a layer of water.
   (*a*) Give the equation describing the formation of 1,2-dibromoethane by the above reaction.
   (*b*) The common laboratory source for the preparation of ethene is ethanol. Briefly explain how ethene may be prepared from ethanol, giving the appropriate equation(s).
   (*c*) Calculate (i) the mass of bromine, (ii) the volume at s.t.p. of ethene, that would theoretically be required for the preparation of 10 g of 1,2-dibromoethane as in (*a*).
   (*d*) Give the equation(s) for the reaction of 1,2-dibromoethane with (i) aqueous potassium hydroxide solution, (ii) alcoholic potassium hydroxide solution.
   (*e*) How may 1,2-dibromoethane be converted into 1,2-diaminoethane? Equation(s) should be given.
   (London, 1972)

4. The elimination of hydrogen bromide from the compound $CH_3CH_2CH(Br)CH_3$ by hot, concentrated, alcoholic potassium hydroxide can lead, in principle, to **two** possible products.
   (*a*) What are the **two** products?
   (*b*) Give **one** method by means of which they could be distinguished from each other.
   (*c*) Which one of the products exists in two stereoisomeric forms? Give a diagram of each form and indicate whether they are optical or geometrical isomers.
   Write down the structure of each possible unsaturated product which could, in principle, be obtained by elimination of hydrogen bromide from the dibromide $CH_3CH(Br)CH(Br)CH_3$.
   (O and C, 1973)

5. Outline reaction schemes, indicating reactants and essential conditions only, by which a chloroalkane may be converted into derivatives where the chlorine atom is replaced by any **three** functional groups, such as hydroxyl, carboxylic acid, amino or any other you may choose.
   Discuss the oxidation of hydroxyalkanes.
   Describe the experimental procedure for any reaction described in this question which you have seen carried out.
   (NISEC, 1974)

6. (*a*) 1-Iodopentane may be hydrolyzed by boiling it with aqueous sodium hydroxide.
   (i) Give the name and the formula of the carbon compound formed in this reaction.
   (ii) The reaction is second order overall and proceeds by a bimolecular mechanism. Suggest a structural formula for the 'activated complex' (transition state).
   (iii) Draw an 'energy profile' for the hydrolysis reaction. The diagram should be fully labelled.

(*b*) Using 1-idopropane ($CH_3CH_2CH_2I$) as *one* starting material, suggest synthetic pathways for the preparation of the following. Clearly indicate any reagents required (organic or otherwise).

(i) Methyl propyl ether, $CH_3CH_2CH_2OCH_3$.

(ii) Butanoic acid, $CH_3CH_2CH_2CO_2H$.

(*c*) Arrange the following halogen compounds in order of the rate at which they may be expected to hydrolyze, beginning with the one having the fastest rate.

A $CH_3CH_2C(=O)Cl$

B C_6H_5–Cl

C C_6H_5–CH_2I

D C_6H_5–CH_2Cl

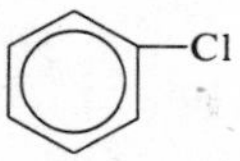

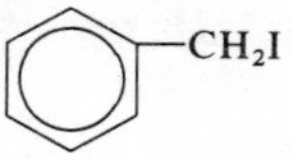

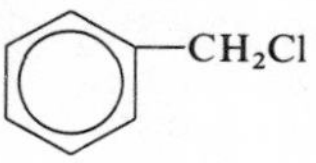

E C_6H_5–CH_2Br

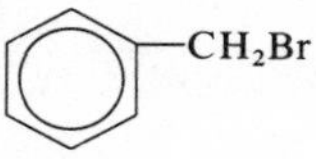

(*d*) (i) Briefly describe a simple experimental procedure that could be used to demonstrate the difference in the rate of hydrolysis between these substances.

(ii) What particular precautions would you advise to ensure that the experiment gives meaningful results?

(London, 1973)

Chapter 9

Hydroxyl and Alkoxyl Functions

Compounds containing hydroxyl functions –OH attached to sp^3 carbon are *alcohols*. When this function is attached to an aromatic nucleus, its properties are considerably modified, and these hydroxybenzenes are called *phenols*. Compounds containing alkoxyl functions –OR are *ethers*.

Nomenclature

The suffix for the hydroxyl group is *-ol*. The following examples make this clear, and also give the trivial names for the simpler members of the series.

CH_3OH
methanol
(methyl alcohol)

CH_3CH_2OH
ethanol
(ethyl alcohol)

$CH_3CH_2CH_2OH$
propan-1-ol
(n-propyl alcohol)
(primary)

$CH_3CH(OH)CH_3$
propan-2-ol
(isopropyl alcohol)
(secondary)

$CH_3CH_2CH_2CH_2OH$
butan-1-ol
(n-butyl alcohol)
(primary)

$CH_3CH_2CH(OH)CH_3$
butan-2-ol
(sec-butyl alcohol)
(secondary)

$(CH_3)_2CHCH_2OH$
2-methylpropan-1-ol
(isobutyl alcohol)
(primary)

$(CH_3)_3COH$
2-methylpropan-2-ol
(t-butyl alcohol)
(tertiary)

The terms 'primary', 'secondary', and 'tertiary' have the same meanings as previously: thus, primary alcohols contain the group $-CH_2OH$, secondary alcohols contain the group $>CHOH$, and tertiary alcohols contain the group $-\overset{|}{C}OH$ (i.e. $\equiv COH$). Names such as 'isopropanol', 't-butanol', etc. are hybrids of the two systems and should not be used; t-butyl alcohol, for example, is not a butanol, but must be systematically named as a propanol. The misnomer 't-butanol' is therefore incorrect and can cause confusion.

For phenols, trivial names are often used. Alternatively they can be named as hydroxy-derivatives of their parent hydrocarbons or, better, as alkyl derivatives of phenol.

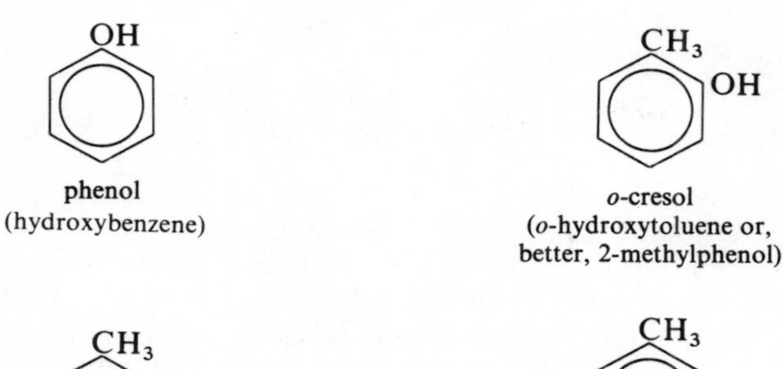

phenol
(hydroxybenzene)

o-cresol
(*o*-hydroxytoluene or, better, 2-methylphenol)

m-cresol
(*m*-hydroxytoluene or 3-methylphenol)

p-cresol
(*p*-hydroxytoluene or 4-methylphenol)

Phenol is the only member of this series which needs to be considered in detail here. The side-chain alcohol $C_6H_5CH_2OH$ is isomeric with the cresols. Its systematic name is phenylmethanol, but it is usually known as benzyl alcohol. It behaves for the most part like other alcohols.

Ethers are named systematically as alkoxy-derivatives of hydrocarbons, but names derived from the alkyl groups attached to the oxygen atom are commonly used for the simpler members; e.g. $CH_3CH_2OCH_2CH_3$, ethoxyethane or diethyl ether; $CH_3OCH_2CH_5$, methoxyethane or methyl ethyl ether.

The most common ether, ethoxyethane, is often called simply 'ether'.

Methods for the introduction of the hydroxyl function

The methods are mentioned briefly here, with cross-references to the detailed discussion of the reactions in their appropriate contexts.

Hydrolysis

The hydrolysis of several types of compounds containing monovalent functions gives hydroxy-compounds. In general, the process is:

$$R{-}X \xrightarrow{OH^-} R{-}OH + X^-$$

The hydrolysis of halides has already been discussed (Chapter 8, pages 77–82), and is an important route to alcohols, although not to phenols.

Esters (cf. Chapter 11, pages 125–7) can also be hydrolyzed to give alcohols, e.g.

$$\underset{\text{ethyl acetate (ethyl ethanoate)}}{CH_3CO.OCH_2CH_3} \xrightarrow{OH^-} CH_3CH_2OH + CH_3CO.O^-$$

This important process, and the conditions under which it occurs, are further discussed in Chapter 11. In particular, alkyl hydrogensulphates, which are esters, can be obtained by addition of sulphuric acid to alkenes, and their hydrolysis gives alcohols; e.g.

$$CH_2{=}CH_2 + H_2SO_4 \longrightarrow CH_3CH_2.O.SO_2OH \xrightarrow{H_2O} CH_3CH_2OH + H_2SO_4$$

Ethanol is prepared thus on an industrial scale.

Esters of phenols can also be hydrolyzed to give phenols, although phenols are more usually prepared by the hydrolysis of salts of arenesulphonic acids, e.g. $ArSO_3^-\ Na^+$, under the vigorous conditions necessary for the approach of the nucleophile to the aromatic nucleus. In practice the reaction is carried out by fusing the sodium arenesulphonate with sodium hydroxide.

$$C_6H_5SO_3^- \xrightarrow[\text{fuse}]{OH^-} C_6H_5O^- + HSO_3^-$$

Thus benzene can be converted into phenol in two stages, by sulphonation (cf. Chapter 7, pages 68–9) followed by this sodium hydroxide fusion. Phenols are themselves weak acids (see page 91) so it is the sodium salt of the phenol which is formed. The free phenol can then be obtained by acidification with mineral acid.

$$C_6H_6 \xrightarrow[\text{reflux}]{\text{conc. } H_2SO_4} C_6H_5SO_3H \xrightarrow{\text{fuse Na salt with NaOH}} C_6H_5ONa \xrightarrow{\text{dilute acid}} C_6H_5OH$$

Arenediazonium salts, $ArN_2^+X^-$ (see Chapter 12, page 140), can also be hydrolyzed to give phenols. Elementary nitrogen is formed in this reaction, which therefore occurs readily, as for example when aqueous solutions of diazonium salts are boiled.

$$C_6H_5N_2^+ \xrightarrow[\text{boil}]{H_2O} C_6H_5OH + H^+ + N_2$$

So great is the tendency of $-N_2^+$ to act as a leaving group that an S_N1 process occurs:

$$ArN_2^+ \underset{}{\overset{\text{slow}}{\rightleftharpoons}} Ar^+ + N_2$$

$$Ar^+ + H_2O \xrightarrow{\text{fast}} ArOH + H^+$$

Nucleophilic substitution at aromatic nuclei is normally difficult and therefore unusual, because the electron-rich nucleus inhibits the approach of the nucleophile. This difficulty is avoided here because the S_N1 mechanism is exceptionally favoured, and the rate-determining stage does not involve the approach of the nucleophile.

Reduction of aldehydes and ketones

Aldehydes and ketones are oxidation products of primary and secondary alcohols, respectively (cf. page 92) and by the reverse process (reduction), alcohols can be obtained (Chapter 10, page 106).

$$\underset{\text{aldehyde}}{RCHO} \xrightarrow{(2H)} \underset{\text{primary alcohol}}{RCH_2OH}$$

$$\underset{\text{ketone}}{RCO.R'} \xrightarrow{(2H)} \underset{\text{secondary alcohol}}{RCH(OH)R'}$$

From Grignard reagents

Primary, secondary, and tertiary alcohols can be obtained by addition of Grignard reagents to appropriate aldehydes and ketones (Chapter 10, page 107). Additionally primary alcohols can be obtained by addition of Grignard reagents to epoxyethane, e.g.

$$CH_3MgI + H_2C{-}CH_2 \text{ (bridged by O)} \longrightarrow CH_3{-}CH_2{-}CH_2{-}OMgI$$

$$\xrightarrow{\text{dilute mineral acid (e.g. HCl)}} \underset{\text{propan-1-ol}}{CH_3CH_2CH_2OH} + MgClI$$

This reaction results in the extension of the carbon chain by two carbon atoms, using the following sequence.

$$ROH \xrightarrow[\text{(cf. page 93)}]{I_2/\text{red P}} RI \xrightarrow[\text{ether}]{\text{Mg/dry}} RMgI \xrightarrow[\text{then dil. acid}]{CH_2{-}CH_2 \text{ (bridged by O)}} RCH_2CH_2OH$$

Oxidation of a Grignard reagent gives the alcohol containing the same number of carbon atoms as the reagent, but since this product can in any case be obtained directly by hydrolysis of the halide from which the Grignard reagent must be prepared, it is not a very useful reaction.

$$RX \xrightarrow[\text{dry ether}]{Mg} RMgX \xrightarrow{O_2} ROMgX \xrightarrow{\text{acid}} ROH$$

$$RX \xrightarrow{OH^-} ROH$$

Fermentation

The classic reaction whereby ethanol is formed is the fermentation of carbohydrates in the presence of enzymes such as those present in yeast (Chapter 13, page 159). The reaction by which ethanol is formed from glucose can be summarized as

$$C_6H_{12}O_6 \longrightarrow 2C_2H_5OH + 2CO_2$$

This process, as well as the hydrolysis of ethyl hydrogensulphate (page 94) is operated on a large scale. Both methods give aqueous solutions of ethanol which are fractionally distilled to give a constant-boiling mixture (an azeotrope) containing 96 per cent ethanol and 4 per cent water (*'rectified spirit'*). Pure ethanol is obtained from this by adding a little benzene and refractionating. The first fraction is a ternary mixture of b.p. 64.8 °C containing benzene, ethanol, and water, and all the water is removed in this. It is followed by a binary mixture of ethanol and benzene, b.p. 68.2 °C, in which all the benzene is removed. Finally pure ethanol, b.p. 78.1 °C, distils. The mixtures contain only relatively small quantities of ethanol, so not much is lost.

Industrial methylated spirit consists of 95 per cent rectified spirit and 5 per cent crude methanol. *Mineralized methylated spirit* is 90 per cent rectified spirit, 9 per cent methanol, a little pyridine, high-boiling petroleum, and a purple dye. These toxic impurities are added to render the products unfit for drinking, so that they may be sold without excise duty being charged.

Occurrence of hydroxy-compounds

Alcohols occur in nature mainly in combination with acids as esters (cf. Chapter 11). Phenols occur in coal tar and this has been, and probably will continue to be, an important source of them. They are obtained from the appropriate fractions by extraction with alkali. However, most phenol is now prepared industrially by autoxidation of cumene (Chapter 5, page 43).

Properties and reactions of alcohols and phenols

The lower alcohols are liquids (methanol, b.p. 64 °C; ethanol, b.p. 78 °C; etc.). Being hydroxy-compounds they have an affinity for water (see page 00). Thus methanol, ethanol, and the propanols are completely miscible with water, but the solubility in water falls off as the series is ascended. Simple phenols are low melting point solids, or involatile liquids. Phenol ('carbolic acid'), for example, melts at 43 °C and boils at 183 °C. Phenols are moderately soluble in water, have characteristic smells, and are used as antiseptics and disinfectants.

Acidity and Basicity

Just as water is amphoteric,

Acid: $$H_2O \rightleftharpoons H^+ + OH^-$$

Base: $$H_2\ddot{O} + H^+ \rightleftharpoons H_3O^+$$

so hydroxy-compounds can in principle act as acids or bases.

Acids: $$ROH \rightleftharpoons RO^- + H^+$$

Bases: $$R\ddot{O}H + H^+ \rightleftharpoons ROH_2^+$$

The acidity or basicity depends largely on the framework to which the hydroxyl group is attached. Thus electron repulsion by alkyl groups renders the oxygen more electronegative and the ability of the unshared pair of electrons to co-ordinate with protons is enhanced.

$$\overset{\delta+}{R} \rightarrow \overset{\delta-}{OH}$$

Alcohols are therefore slightly stronger bases than water and the oxonium ions, ROH_2^+, are readily formed as intermediates in reactions of alcohols which are catalyzed by strong acids.

On the other hand the ability of a conjugated framework like phenyl to delocalize the negative charge in the anions ArO^- by the resonance:

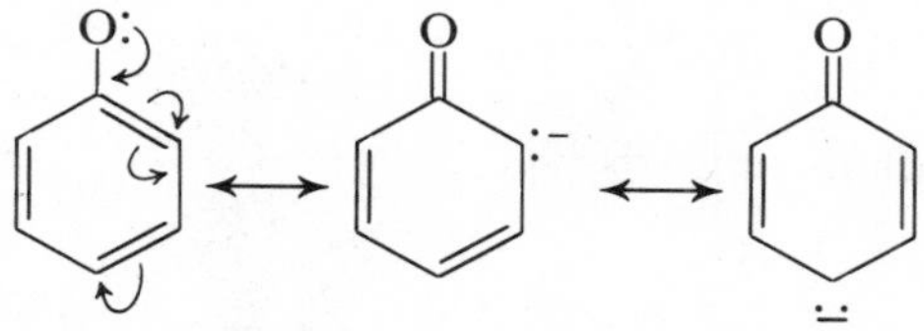

results in considerable stabilization of these anions. (In these formulae the curved arrows show how the electron-pair must be regarded as moving to give the other canonical structures.) The equilibrium:

$$ArOH \rightleftharpoons ArO^- + H^+$$

is therefore much further to the right than the corresponding dissociation of alcohols, and phenols behave like very weak acids, (weaker for example than carbonic acid). The value of pK_a ($-\log K_a$, where K_a is the equilibrium constant for the dissociation) for phenol is 9.98. Thus, although they will dissolve in sodium hydroxide to give salts:

$$PhOH + NaOH \longrightarrow \underset{\text{sodium phenoxide}}{PhO^-Na^+} + H_2O,$$

phenols do not displace carbon dioxide from carbonates.

Although alcohols are even weaker acids than phenols, they do display their acidic as well as their basic properties in some reactions. Thus solutions of caustic alkalis in them contain alkoxide ions. For example, a solution of potassium hydroxide in ethanol ('alcoholic potash': sodium hydroxide is insoluble in ethanol) contains ethoxide ions.

$$EtOH + OH^- \rightleftharpoons H_2O + EtO^-$$

The strong basicity of alkoxide ions has already been noted (Chapter 8, page 83), and it arises because alcohols are very weak acids indeed. Moreover, alkali metals react with alcohols to give the alkoxides and hydrogen, although the reactions are less vigorous than the corresponding reactions with water.

$$2EtOH + 2Na \longrightarrow \underset{\text{sodium ethoxide}}{2EtONa} + H_2$$

Thus alcohols, like water, are amphoteric. They are, however, stronger bases but weaker acids than water.

Oxidation

Primary, secondary, and tertiary alcohols behave differently on oxidation, and the different reactions can be used to distinguish them.

With mild oxidizing agents, such as dichromate/sulphuric acid, *primary* alcohols give *aldehydes*.

$$RCH_2OH \xrightarrow{(O)} \underset{\text{an aldehyde}}{RC(=O)H} + H_2O$$

Secondary alcohols similarly give *ketones*.

$$RR'CHOH \xrightarrow{(O)} \underset{\text{a ketone}}{RR'C{=}O} + H_2O$$

Both aldehydes and ketones contain the *carbonyl* function $>C{=}O$, which is the subject of Chapter 10. Other oxidizing agents can also be used, but need not be discussed here.

Tertiary alcohols are not oxidized by such mild reagents, but with very vigorous oxidizing agents the carbon framework breaks up, giving complex mixtures of products (this, however, also occurs with most organic compounds on such vigorous oxidation).

Halogenation

Hydroxyl groups attached to sp^3 carbon can be replaced by the halogens. The halogen acids themselves can accomplish this and the order of their reactivity is HI > HBr > HCl. The acidic nature of these reagents is essential because this, like other reactions in which the hydroxyl group is replaced, is catalyzed by strong acids. The acid catalyst works by protonating the oxygen:

$$ROH + H^+ \rightleftharpoons R\overset{+}{O}H_2$$

Nucleophilic replacement (by the halide anion in this case) is much easier with this oxonium salt (the *conjugate acid* of the alcohol):

$$X^- + R\overset{+}{O}H_2 \longrightarrow XR + H_2O$$

than with the alcohol itself:

$$X^- + ROH \longrightarrow XR + \bar{O}H$$

because the water molecule is a much better leaving group than the hydroxide ion on account of its greater stability.

The actual replacement process on the conjugate acid is mechanistically analogous to other nucleophilic replacements, such as those of halides discussed in Chapter 8. Thus its mechanism is usually S_N2 with primary alcohols:

$$\ddot{X}^- + R{-}\overset{+}{O}H_2 \longrightarrow \overset{\delta-}{X}\cdots R\cdots \overset{\delta+}{O}H_2 \longrightarrow XR + H_2O$$

and S_N1 with tertiary alcohols:

$$R{-}\overset{+}{O}H_2 \xrightleftharpoons{\text{slow}} R^+ + H_2O$$

$$R^+ + X^- \xrightarrow{\text{fast}} RX$$

Secondary alcohols can react by either mechanism, depending on the conditions.

Other halogenating agents are commonly used, notably phosphorus trihalides, phosphorus pentahalides, and sulphur dichloride oxide (thionyl chloride). Mixtures of bromine or iodine with red phosphorus are effective sources of the phosphorus halides *in situ* for bromination or iodination. The reactions proceed as follows:

$$3ROH + PX_3 \longrightarrow 3RX + \underset{\text{phosphonic acid}}{P(OH)_3[H_3PO_3]}$$

$$ROH + PX_5 \longrightarrow RX + HX + \underset{\text{phosphorus trihalide oxides}}{POX_3}$$

The evolution of hydrogen chloride gas with phosphorus pentachloride in the cold is regarded as a test for the hydroxyl group. The reaction with sulphur dichloride oxide as follows:

$$ROH + SOCl_2 \longrightarrow RCl + HCl + SO_2$$

Hydroxyl groups in phenols cannot readily be replaced by halogens because of the difficulty of effecting nucleophilic replacement at sp^2 carbon. Thus phenol gives only a very poor yield of chlorobenzene when treated with phosphorus pentachloride.

Alkylation

Alkoxide and phenoxide ions are good nucleophiles, and can therefore replace halogens in alkyl halides as discussed in Chapter 8, page 78. The iodides are usually used. This results in the formation of ethers, and amounts formally to alkylation of the hydroxyl group (i.e. replacement of its hydrogen by an alkyl group).

$$RO^- + R'X \longrightarrow ROR' + X^-$$

Thus treatment of sodium ethoxide (prepared by dissolving sodium in ethanol) with iodoethane gives diethyl ether (ethoxyethane):

$$C_2H_5O^-Na^+ + C_2H_5I \longrightarrow C_2H_5OC_2H_5 + Na^+I^-$$

This is the *Williamson ether synthesis*.

Similarly, treatment of sodium phenoxide with iodomethane gives methoxybenzene, which is usually known by its trivial name 'anisole'.

$$C_6H_5O^-Na^+ + CH_3I \longrightarrow C_6H_5OCH_3 + Na^+I$$

methoxybenzene
(anisole)

These reactions are nucleophilic substitutions and can proceed by S_N1 or S_N2 mechanisms depending mainly on the alkyl groups in the alkylating agents (cf. Chapter 8, page 81).

The dialkyl sulphates, e.g. dimethyl sulphate, $(CH_3)_2SO_4$, can be used instead of the iodides as alkylating agents.

Ethyl hydrogensulphate is the alkylating agent in the 'continuous etherification process' by which diethyl ether is manufactured and can also be prepared in the laboratory. Ethanol and concentrated sulphuric acid are heated at 140 °C to give ethyl hydrogensulphate, and more ethanol is added gradually. The volatile ether (b.p. 34.5 °C) distils off continuously.

$$EtOH + H_2SO_4 \longrightarrow EtHSO_4 + H_2O$$

$$EtHSO_4 + EtOH \longrightarrow EtOEt + H_2SO_4$$

The temperature must not be allowed to rise above about 140 °C to avoid formation of ethane (cf. Chapter 6, page 46). Sulphuric acid is regenerated in the second stage and is available to react with more ethanol. Thus large amounts of ether can be synthesized before the formation of by-products and the accumulation of impurities make it necessary to add fresh sulphuric acid.

The nucleophile in this reaction is the ethanol molecule rather than the ethoxide ion, because the medium is strongly acidic. Although some ethanol molecules are present as the conjugate acid, the reaction

$$EtOH + H^+ \rightleftharpoons Et\overset{+}{O}H_2$$

is reversible and sufficient unprotonated molecules are present to function as the nucleophiles in this reaction.

Esterification

Alcohols react with organic and inorganic acids in the presence of strong acid catalysts (usually concentrated sulphuric acid) to give *esters*. For example, the ester ethyl ethanoate (ethyl acetate) is formed when ethanol, ethanoic acid (acetic acid), and concentrated sulphuric acid are heated together.

$$\underset{\substack{\text{ethanoic acid}\\\text{(acetic acid)}}}{CH_3CO.OH} + \underset{\text{ethanol}}{CH_3CH_2OH} \xrightleftharpoons{H_2SO_4} \underset{\substack{\text{ethyl ethanoate}\\\text{(ethyl acetate)}}}{CH_3CO.OCH_2CH_3} + H_2O$$

This, like all esterifications by this method, is a reversible reaction. The volatile ester is therefore continuously distilled off to disturb the equilibrium and make the reaction continue to proceed in the forward direction.

The function of the sulphuric acid is to act as a catalyst. From the earlier discussions it might be guessed that, since the sulphuric acid protonates the alcohol, the mechanism of this reaction might be analogous to those discussed earlier, namely

$$EtOH + H^+ \rightleftharpoons Et\overset{+}{O}H_2$$

$$CH_3CO.O^- + Et\overset{+}{O}H_2 \longrightarrow CH_3CO.OEt + H_2O \quad (S_N1 \text{ or } S_N2)$$

This is, indeed a possible mechanism for esterification, although there are others which are more common. In fact, because of the properties of the –CO·OH group in the organic acids, esterification is rather a complex reaction mechanistically. Some further discussion is to be found in Chapter 11 (page 116).

Dry hydrogen chloride, passed in as the gas until its concentration is 3–5 per cent, can be used in place of sulphuric acid as the acid catalyst for esterification. This is called the *Fischer–Speier* method.

Esters can also be prepared by treating alcohols with acyl halides (usually chlorides). These acyl halides are regarded as derivatives of the acids (like acetic acid), and are prepared from them (see Chapter 11, page 116). For example acetyl chloride (ethanoyl chloride) can be used to prepare acetate (ethanoate) esters.

$$ROH + \underset{\text{acetyl chloride}}{CH_3CO.Cl} \longrightarrow \underset{\text{alkyl acetate}}{CH_3CO.OR} + HCl$$

Acid anhydrides (cf. Chapter 11, page 116) such as acetic (ethanoic) anhydride can be used similarly.

$$ROH + \underset{\text{acetic anhydride}}{(CH_3CO)_2O} \longrightarrow \underset{\text{alkyl acetate}}{CH_3CO.OR} + \underset{\text{acetic acid}}{CH_3CO.OH}$$

These methods are useful because the reactions are not reversible. They can also be used to esterify phenols, which are difficult to esterify with the acids themselves, e.g.

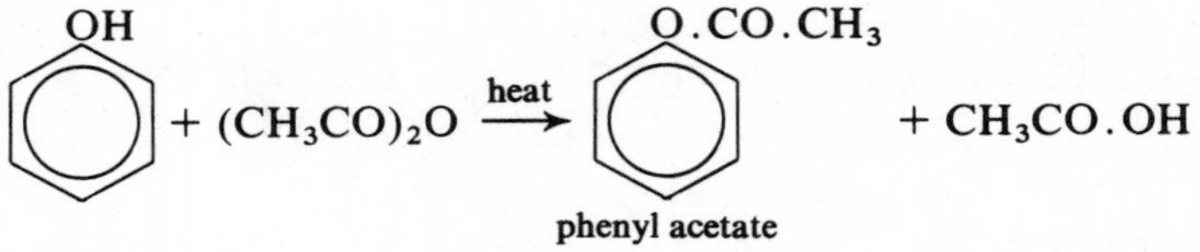

$$C_6H_5OH + (CH_3CO)_2O \xrightarrow{\text{heat}} C_6H_5O.CO.CH_3 + CH_3CO.OH$$

phenyl acetate

Dehydration

The dehydration of alcohols to give alkenes has already been dealt with (Chapter 6, page 46).

Reactions of the nuclei of phenols

Because the oxygen atom has two pairs of unshared electrons, these can be delocalized into the nucleus. The hydroxyl group in phenols therefore behaves as an electron-repelling group and the nucleus becomes richer in electron density. It is therefore much more reactive towards electrophiles than is benzene (cf. Chapter 7, page 67). The positions to which the electrons can be repelled are shown by the curved arrows below.

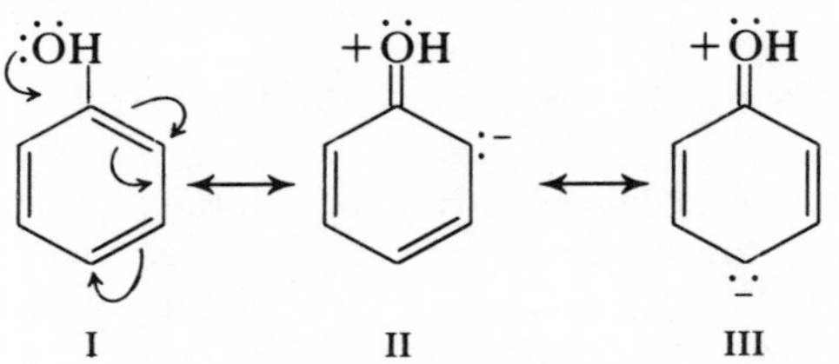

:ÖH　　+ÖH　　+ÖH

I　　II　　III

This delocalization is very slight in phenol itself; the contributions of forms II and III to the resonance hybrid are very small—much smaller than the corresponding forms in the phenoxide ion—because II and III are dipoles and hence a good deal more energetic than I, which is neutral. In the phenoxide ion, on the other hand all the forms are negatively charged and the only difference is the position of the negative charge.

Thus, as explained in Chapter 7, page 67, the result is that the 2-(*ortho*-) and 4-(*para*-) positions become richer in electron density and hence more liable to attack by electrophiles. This does not happen at the 3-(*meta*-) positions because this would require the participation of a canonical structure like IV, which contains a bridging bond and is a highly unstable, strained, bicyclic (i.e. 'two ringed') structure.

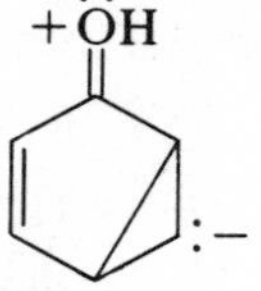

+ÖH

IV

Being much more energetic than II and III the contribution of IV to the hybrid is negligible.

Thus it is the *ortho*- and *para*- positions which are '*activated*' to electrophilic substitution by the hydroxyl group. These substitutions are many orders of magnitude more rapid than they are in benzene; for example phenol is brominated in the two *ortho*-positions (2- and 6-), and the one *para*-position (4-) with bromine water at room temperature.

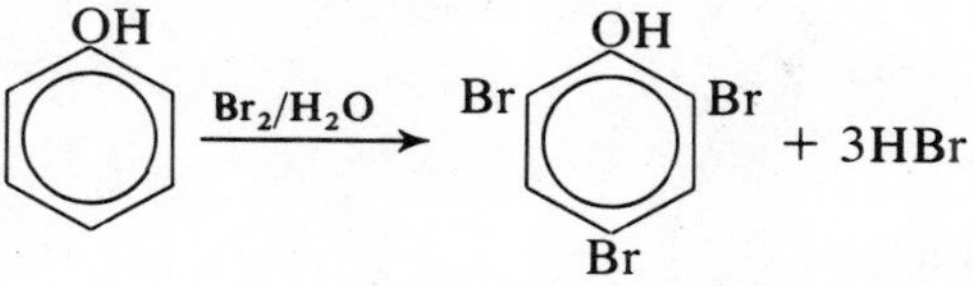

$$C_6H_5OH \xrightarrow{Br_2/H_2O} C_6H_2Br_3OH + 3HBr$$

2,4,6-tribromophenol

The product is a pale yellow precipitate which is formed immediately. Nitration of phenol is similarly easy and leads ultimately to the trinitro-compound picric acid (2,4,6-trinitrophenol).

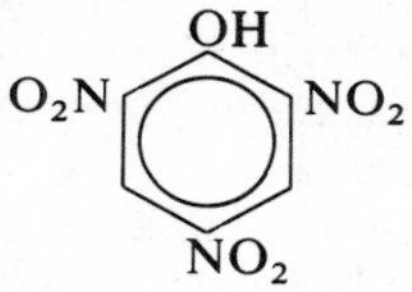

$$C_6H_2(NO_2)_3OH$$

picric acid

Polyhydroxy-compounds

Compounds containing two hydroxyl groups are known as dihydric alcohols or dihydric phenols, those with three hydroxyl groups trihydric, and so on. The simplest dihydric alcohol is ethane-1,2-diol (ethylene glycol), and the simplest trihydric alcohol is propane-1,2,3-triol (glycerol).

$$CH_2OH{-}CH_2OH \qquad CH_2OH{-}CHOH{-}CH_2OH$$

ethane-1,2-diol (ethylene glycol) — propane-1,2,3-triol (glycerol)

Ethane-1,2-diol can be prepared by hydrolysis of 1,2-dibromoethane:

$$CH_2Br{-}CH_2Br \xrightarrow{2OH^-} CH_2OH{-}CH_2OH + 2Br^-$$

or of epoxyethane with dilute hydrochloric acid:

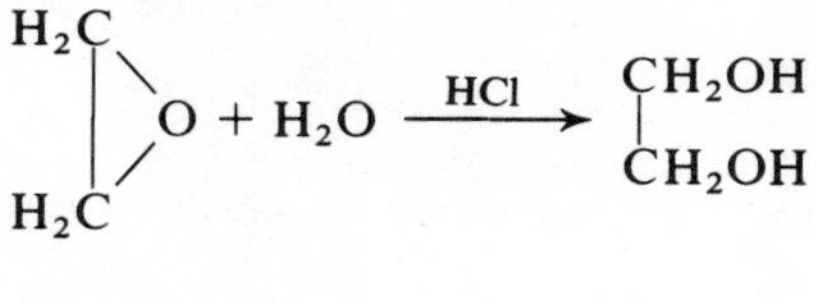

$$H_2C{-}CH_2(O) + H_2O \xrightarrow{HCl} CH_2OH{-}CH_2OH$$

The latter method is used industrially, the epoxyethane being obtained by oxidation of ethene with oxygen under pressure in the presence of a catalyst.

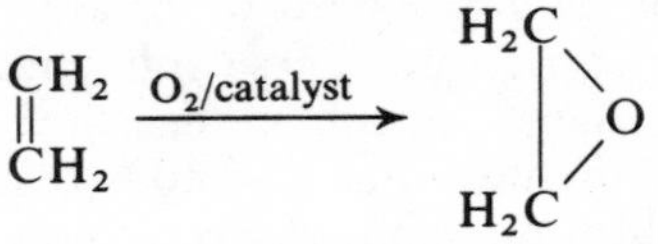

$$\begin{matrix} CH_2 \\ \| \\ CH_2 \end{matrix} \xrightarrow{O_2/\text{catalyst}} \begin{matrix} H_2C \\ | \\ H_2C \end{matrix}\!\!>\!O$$

Ethane-1,2-diol is also formed when ethene is oxidized with cold dilute alkaline permanganate.

$$\begin{matrix} CH_2 \\ \| \\ CH_2 \end{matrix} + H_2O + (O) \xrightarrow[0\ ^\circ C]{\text{alkaline } KMnO_4} \begin{matrix} CH_2OH \\ | \\ CH_2OH \end{matrix}$$

It is a viscous, colourless liquid, b.p. 197 °C, completely miscible with water. It dissolves many inorganic salts, and is widely used as an antifreeze. It behaves chemically as expected, i.e. it displays the expected reactions of the two hydroxyl groups.

Propane-1,2,3-triol (glycerol, 'glycerine') occurs naturally as its esters with some long chain organic acids in oils and fats. It is obtained by hydrolysis of these esters in soap manufacture (see Chapter 11, page 126), and is usually seen as a viscous colourless liquid (m.p. 18 °C, b.p. 290 °C). It is hygroscopic and has a sweet taste. Because of the abundance of hydrophilic ('water-loving') hydroxyl groups, which form hydrogen bonds with water molecules, it is extensively solvated by water, and hence very soluble in water.

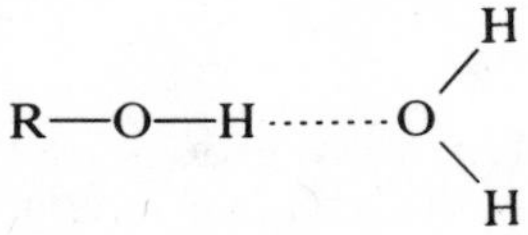

$$R{-}O{-}H \cdots\cdots O\begin{matrix} \diagup H \\ \diagdown H \end{matrix}$$

This is the reason for the water-solubility of all hydroxy-compounds, and glycerol shows this behaviour very strongly. Hence it is completely miscible with water but insoluble in solvents like ether or benzene, which interact most strongly with the organic, hydrocarbon parts of solute molecules.

Ethers (alkoxy-compounds)

Ethers, which are prepared by alkylation of hydroxy-compounds, are isomeric with alcohols and with one another. For example, the following compounds are all isomeric.

C_4H_9OH
the four 'butyl' alcohols

$CH_3CH_2OCH_2CH_3$
ethoxyethane
(diethyl ether)

$CH_3OCH_2CH_2CH_3$
1-methoxypropane
(methyl n-propyl ether)

$CH_3OCH(CH_3)_2$
2-methoxypropane
(methyl isopropyl ether)

The isomeric relationship of ethoxyethane with the methoxypropanes, which is a result of the presence of two alkyl groups, is called '*metamerism*'.

Ethers are generally more volatile than the corresponding alcohols, because, since they contain no hydroxyl groups, hydrogen bonding is much less prevalent. Thus ethoxyethane (diethyl ether) boils at 34.5 °C. Its vapour is highly flammable and forms explosive mixtures with air. It should therefore be handled with great caution. It has been used as an anaesthetic, and as a solvent.

Basicity

The simpler ethers are somewhat miscible with water, although much less so than alcohols. Ethoxyethane, for example, at 20 °C dissolves 1.5 per cent by weight of water, while at the same temperature water dissolves 6.5 per cent of ethoxyethane. Ethers are, however, much more soluble in strong acids, because of their basicity. They form oxonium salts, e.g. with hydrochloric acid:

$$C_2H_5\ddot{O}C_2H_5 + H^+Cl^- \rightleftharpoons (C_2H_5)_2\overset{+}{O}{-}H \quad Cl^-$$

Ethers are stronger bases than alcohols because they contain two alkyl groups, both of which repel electrons towards the oxygen atom by their $+I$ effects.

$$R{\rightarrow}\ddot{\underset{\cdot\cdot}{O}}{\leftarrow}R$$

The oxygen thus becomes more electronegative and the unshared pairs of electrons more prone to co-ordinate with the proton, i.e. the basicity is increased.

Aromatic ethers (alkoxyarenes) are not so basic, however, because the unshared pairs of electrons can be delocalized in the nucleus, and hence rendered less available at the oxygen for co-ordination. Alkoxy-groups, like hydroxyl groups, therefore exert a conjugative effect. It also follows, therefore, that alkoxy-groups attached to aromatic nuclei are activating and *ortho-para*-directing towards electrophilic substitution in those nuclei (cf. Chapter 7, pages 66–8).

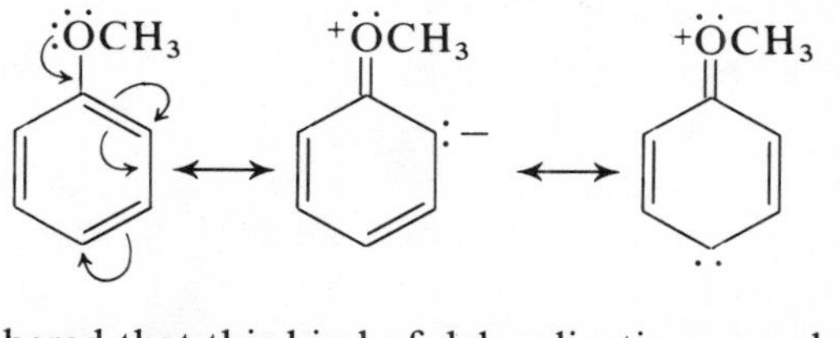

$:\ddot{O}CH_3$ $\longleftrightarrow$ $^+\ddot{O}CH_3$ $\longleftrightarrow$ $^+\ddot{O}CH_3$

It will be remembered that this kind of delocalization was also responsible for the enhanced acidity of the hydroxyl group in phenols, and it will be seen later (Chapter 12, page 140) that similar delocalization of unshared electrons on nitrogen is the reason why the basicities of aminoarenes are lower than those of aminoalkenes.

Reactions

Ethers are much less reactive than alcohols, because of the absence of the hydrogen of the hydroxyl group. Thus they do not react with sodium or with phosphorus pentachloride in the cold. They do however give iodo-compounds on heating with concentrated hydroiodic acid.

$$EtOEt + 2HI \longrightarrow 2EtI + H_2O$$

Questions

1. Describe in outline **one** process for the manufacture of phenol.

   Give reagents, reaction conditions and equations to show how you would convert phenol into

   (*a*) OCH_3 on a benzene ring (methoxybenzene) (*b*) OH on a benzene ring with NO_2 para (*c*) $OCO.C_6H_5$ on a benzene ring

   Explain why (i) phenol is mildly acidic but ethanol is not, (ii) 4-nitrophenol is much more acidic than phenol.

   (AEB, 1973)

2. (*a*) Give the names and structural formulae of **three** alcohols having the molecular formula C_4H_9OH. In each case classify the alcohol as primary, secondary or tertiary.

   (*b*) By means of equations and brief notes of experimental conditions, state how an aliphatic primary alcohol may be converted to: (i) an aldehyde, (ii) an ester, (iii) an amine.

   (*c*) Outline an industrial procedure for the production of ethanol (ethyl alcohol). Indicate how the last traces of water can be removed from the product.

   (JMB, 1973)

3. Write down **two** reaction schemes which show how benzene may be converted into phenol in the laboratory.

   Describe in outline one **modern** method which is used to prepare **either** phenol **or** ethanol in industry.

   Give **two** reactions in which phenol and ethanol show similar properties and **two** in which they differ.

   (O and C, 1973)

4. The reactivity of a functional group is influenced by neighbouring groups and by its position in the molecule. Give examples to illustrate that this applies to the hydroxyl group (–OH) in organic compounds.

   (AEB, 1972)

Chapter 10

The Carbonyl Function

The carbonyl function is the divalent group $\rangle C{=}O$. If one of the valencies is attached to hydrogen the compound is an *aldehyde* $R{-}C(H){=}O$, RCHO (or ArCHO). If both valencies are attached to organic frameworks, the compound is a *ketone*, $\mathrm{R(R')C=O}$, RCO.R′, (or ArCO.Ar′, or ArCO.R). Ketones can be *simple* (R_2CO) or *mixed* (RCO·R′). The lower members of both series are volatile liquids, completely miscible with water.

The carbon atom of the carbonyl group is sp^2 hybridized, as in alkenes. The double bond likewise consists of a σ bond involving one of the sp^2 hybrid orbitals and a π bond formed by 'sideways' overlap of p_z orbitals on the carbon and the oxygen. The σ bonds formed by the carbon therefore lie in a plane, mutually at about 120° to one another, and there is π electron density at the C=O bond above and below the plane of the molecule:

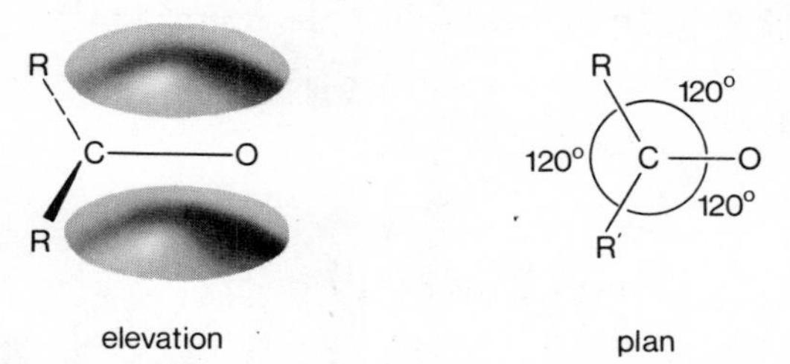

The π electron density is, however, not evenly distributed, but is centred more on the oxygen than on the carbon (see page 104).

Nomenclature

Aldehydes can be given trivial names after the acids to which they can be oxidized. Systematically they are named after the hydrocarbon with the same number of carbon atoms, with the –CHO group indicated by the suffix -al, e.g.

HCHO	CH_3CHO	CH_3CH_2CHO
methanal	ethanal	propanal
(formaldehyde)	(acetaldehyde)	(propionaldehyde)
(b.p. 20 °C)	(b.p. 21 °C)	(b.p. 50 °C)

The trivial names 'formaldehyde' and 'acetaldehyde' are particularly widely used.

Aromatic aldehydes are named systematically by adding the suffix 'carbaldehyde' to the name of the area. Thus the simplest member, C_6H_5CHO, is benzenecarbaldehyde. It is, however, usually known as benzaldehyde.

For systematic naming of ketones, the suffix *-one* is used and the position of the carbonyl group is indicated, if necessary, by a number in the usual way. Sometimes they are named by the groups attached to the carbonyl group, and for the first member of the series, the trivial name 'acetone' is generally used. The following examples illustrate this.

$CH_3CO.CH_3$
propanone
(dimethyl ketone, acetone)
(b.p. 55 °C)

$CH_3CO.CH_2CH_3$
butanone
(methyl ethyl ketone)
(b.p. 80 °C)

$CH_3CH_2CO.CH_2CH_3$
pentan-3-one
(diethyl ketone)
(b.p. 102 °C)

$CH_3CO.CH_2CH_2CH_3$
pentan-2-one
(methyl n-propyl ketone)
(b.p. 102° C)

The last two ketones are metamers. Ketones are also metameric with aldehydes, e.g. propanal and propanone (acetone) are metamers.

The fully aromatic diphenyl ketone (systematic name diphenylmethanone):

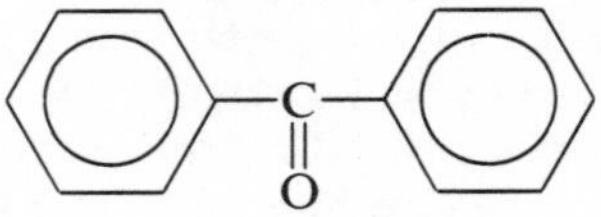

$C_6H_5-C(=O)-C_6H_5$

is usually known by its trivial name benzophenone (m.p. 49 °C), and the mixed methyl phenyl ketone (systematic name phenylethanone):

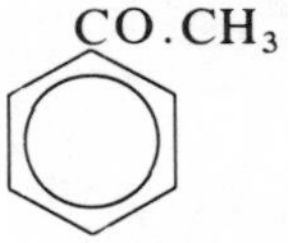

$C_6H_5-CO.CH_3$

is usually known as acetophenone (m.p. 20 °C).

In writing linear formulae for ketones, it is suggested that there is an advantage in following the carbonyl group with a full stop because the oxygen is not in the chain, i.e. it is not bound to the *next* carbon atom. The full stop makes this clear. Thus, for example, $CH_3CO.CH_3$ is clearer than CH_3COCH_3, since the latter may imply the incorrect, and indeed impossible, structure $CH_3C–O–CH_3$.

Formation of the carbonyl group

There are many reactions in which carbonyl groups are formed and therefore many methods for the preparation of aldehydes and ketones, some of them rather complex. The following account therefore includes only the simpler (not necessarily the best) methods.

Oxidation

Aldehydes are formed by oxidation of primary alcohols, and ketones by oxidation of secondary alcohols (Chapter 9, page 92).

$$RCH_2OH \xrightarrow{(O)} RCHO + H_2O$$

$$R(R')CHOH \xrightarrow{(O)} R(R')CO + H_2O$$

Aldehydes can be further oxidized quite readily to carboxylic acids, so the oxidizing agent must be carefully chosen and the conditions carefully controlled to avoid this (cf. page 105). Acid dichromate is often used as the oxidizing agent. Ketones are not readily oxidized so the problem does not arise so acutely with them.

Aldehydes cannot be prepared by direct reduction of carboxylic acids because a vigorous reducing agent is required (Chapter 11, page 123) and further reduction of the aldehyde to a primary alcohol always occurs. They can however be prepared by catalytic hydrogenation of acid chlorides (the 'Rosenmund' reaction, cf. Chapter 11, page 128).

Ozonolysis

Carbonyl compounds are formed by ozonolysis of alkenes (Chapter 6, page 54).

Hydrolysis

Hydrolysis of *gem*-dihalides gives dihydroxy-compounds in which both hydroxyl groups are attached to the same carbon atom. Such compounds are normally too unstable to exist, and lose water immediately on formation to give carbonyl compounds.

$$\underset{\substack{\text{1,1-dichloroethane} \\ \text{(ethylidene chloride)}}}{CH_3CHCl_2} \longrightarrow [CH_3CH(OH)_2] \longrightarrow \underset{\substack{\text{ethanal} \\ \text{(acetaldehyde)}}}{CH_3CHO} + H_2O$$

$$\underset{\text{2,2-dichloropropane}}{CH_3CCl_2CH_3} \longrightarrow [CH_3CH(OH)_2CH_3] \longrightarrow \underset{\text{acetone}}{CH_3CO.CH_3} + H_2O$$

Decomposition of calcium salts

Ketones are obtained when solid calcium salts of carboxylic acids (cf. Chapter 11) are heated. The ketone distils off but the yields are generally rather poor, e.g.

$$(CH_3CO_2)_2Ca \xrightarrow{\text{heat}} CH_3CO.CH_3 + CaCO_3$$

calcium acetate (calcium ethanoate) — acetone — calcium carbonate

If a mixture of two different calcium salts is heated, a mixture of the mixed ketone and both simple ketones is obtained.

If one of the calcium salts is calcium formate (calcium methanoate), an aldehyde results, e.g.

$$(CH_3CO_2)_2Ca + (HCO_2)_2Ca \longrightarrow 2CH_3CHO + 2CaCO_3$$

calcium acetate — calcium formate — ethanal (acetaldehyde)

Other products, including the ketone (in this case acetone) are also formed, so the yield is not very good.

Special methods

The formation of ethanal by passing ethyne into sulphuric acid containing mercury(II) sulphate (Chapter 6, page 55), and the formation of acetone together with phenol by hydrolysis of cumene hydroperoxide (Chapter 5, page 43) are of great industrial importance.

Friedel–Crafts reaction for aromatic ketones

The Friedel–Crafts reaction with acid (acyl) halides is useful for the preparation of aromatic ketones (Chapter 7, page 70); e.g.

$$C_6H_6 + CH_3CO.Cl \xrightarrow{AlCl_3} C_6H_5CO.CH_3$$

phenylethanone (acetophenone)

Reactions of carbonyl compounds

The carbonyl group is highly reactive, and its reactions give such a variety of products that they are central to synthetic organic chemistry. The key to many of these reactions is that, because of the greater electronegativity of oxygen compared with that of carbon, the carbonyl group is *always* polarized in the sense:

$$\gt \overset{\delta+}{C} = \overset{\delta-}{O}$$

Thus it can be considered that the 'centre of gravity' of the π cloud (cf. page 101) is nearer the oxygen than the carbon.

Alternatively, it can be thought of as a resonance hybrid:

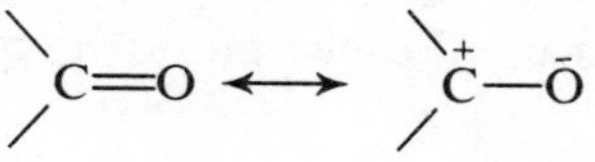

$$\gt C{=}O \longleftrightarrow \gt \overset{+}{C}{-}\bar{O}$$

I II

Since it is dipolar, the contribution of form II is much smaller than that of form I, but nevertheless the carbonyl group is always polarized in this way, and the carbon is therefore susceptible to attack by nucleophiles. This is the essential feature of many carbonyl reactions.

This polarization is even more highly developed in aromatic carbonyl compounds, because aryl nuclei can act as electron sources by virtue of their π electrons.

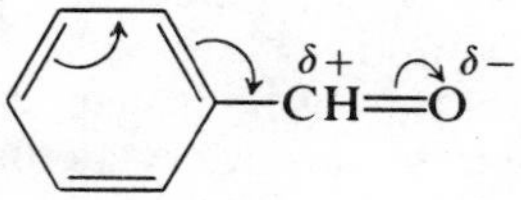

$$C_6H_5{-}\overset{\delta+}{C}H{=}\overset{\delta-}{O}$$

Oxidation

Ketones are not readily oxidized. Vigorous oxidation of them however gives mixtures of carboxylic acids, each containing *fewer* carbon atoms than the ketone and resulting from cleavage of the carbon chain.

Aldehydes are readily oxidized to carboxylic acids containing the *same* number of carbon atoms as the aldehyde.

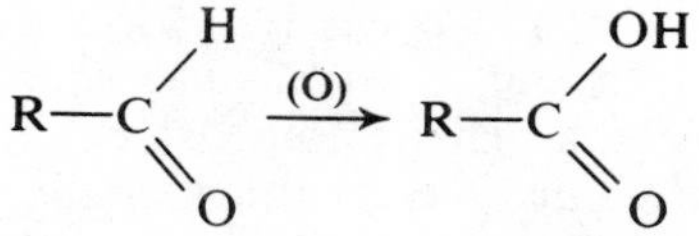

$$R{-}C(H){=}O \xrightarrow{(O)} R{-}C(OH){=}O$$

carboxylic acid

Quite mild oxidizing agents can accomplish this. Acid dichromate is suitable and the reaction by which aldehydes are prepared by oxidation of primary alcohols with this reagent:

$$R{-}CH_2OH \xrightarrow{(O)} RCHO \xrightarrow{(O)} RCO_2H$$

can be stopped at the aldehyde stage only by distilling out the volatile aldehyde as it is formed, and before it can be oxidized further.

In this way aldehydes can be distinguished from ketones, and primary from secondary alcohols. Tertiary alcohols, it will be remembered, are not readily oxidized at all (Chapter 9, page 92).

Because of their ready oxidation, aldehydes are *strong reducing agents*, and this is the basis of distinctive tests which are not undergone by ketones. Thus aldehydes reduce *Fehling's solution* (an alkaline solution containing a complex of copper tartrate), giving a precipitate of red copper(I) oxide on warming. The Fehling's solution acts as if it contained copper(II) oxide.

$$RCHO + 2CuO \longrightarrow RCO_2H + Cu_2O$$

Aldehydes also reduce *Tollen's reagent* (prepared by adding ammonia to silver nitrate solution until the precipitated silver oxide just redissolves). The product is silver which is formed on warming as a silver mirror on the walls of the vessel.

$$RCHO + Ag_2O \longrightarrow RCO_2H + 2Ag$$

Another distinguishing test is that aldehydes restore the magenta colour to *Schiff's reagent* (a solution of the dye rosaniline which has been decolourized with sulphur dioxide). The reactions involved are complex, but the test is simple to carry out in practice.

Addition reactions

Being unsaturated, carbonyl compounds are characterized by addition reactions to the C=O double bond.

Reduction is essentially addition of hydrogen; aldehydes can be reduced to primary alcohols, and ketones to secondary alcohols.

$$RCH{=}O \xrightarrow{(2H)} RCH_2OH$$

$$RR'C{=}O \xrightarrow{(2H)} RR'CHOH$$

Many reagents can be used to bring about these reductions, e.g. Zn/HCl, hydrogen with a platinum or nickel catalyst, sodium tetrahydridoborate(III) (borohydride) ($NaBH_4$), lithium(I) tetrahydridoaluminate(III) (lithium aluminium hydride) ($LiAlH_4$). The tetrahydridoaluminate(III) anion is a nucleophile and the reaction involves its attack on the electropositive carbonyl carbon.

$$\overset{\delta+}{C}{=}\overset{\delta-}{O} \;\; (H{-}\bar{A}lH_3) \longrightarrow H{-}C{-}O^- + AlH_3 \xrightarrow{(H^+)} CH{-}OH$$

The details of the mechanism whereby the oxygen subsequently 'picks up' a proton are complex and need not concern us. The essential feature of the process is, however, the nucleophilic attack by AlH_4^- (and similarly BH_4^-) on the carbonyl carbon.

Aromatic carbonyl compounds can readily be reduced further (cf. Chapter 7, page 70).

$$>C{=}O \longrightarrow >CH_2$$

Other products can be obtained by reduction of carbonyl compounds under special conditions, but these reactions are beyond the scope of this book.

The **addition of Grignard reagents** to carbonyl compounds to give alcohols has been mentioned (Chapter 9, page 89). The reaction is easily rationalized in terms of the polar natures of the reagents as follows.

$$\begin{array}{c}\diagdown \overset{\delta+}{C}{=}\overset{\delta-}{O} \\ \diagup \\ R^- \quad MgX^+\end{array} \longrightarrow \begin{array}{c}\diagdown C{-}\bar{O} \\ \diagup | \\ R \quad MgX^+\end{array} \longrightarrow \begin{array}{c}\diagdown C{-}OMgX \\ \diagup | \\ R\end{array}$$

It involves nucleophilic attack by the carbanion R^- on the somewhat electropositive carbonyl carbon, and this is probably the rate determining stage. Thus this, like the other additions discussed below, is essentially a *nucleophilic addition* (contrast the electrophilic and homolytic additions to alkenes and alkynes). Most of these reactions with carbonyl compounds follow similar patterns.

The kind of alcohol ultimately obtained is determined by the kind of carbonyl compound used. *Methanal* (formaldehyde) with Grignard reagents gives *primary alcohols*.

$$RMgX + H_2C{=}O \longrightarrow H_2\underset{R}{\underset{|}{C}}{-}OMgX \xrightarrow[\text{(e.g. dil. HCl)}]{\text{hydrolysis}} RCH_2OH + MgXCl$$

Other aldehydes give *secondary alcohols*.

$$RMgX + R'CH{=}O \longrightarrow R'\underset{R}{\underset{|}{C}}H{-}OMgX \longrightarrow \begin{array}{c}R \\ \diagdown \\ CHOH \\ \diagup \\ R'\end{array}$$

Ketones give *tertiary alcohols*.

$$RMgX + \begin{array}{c}R' \\ \diagdown \\ C{=}O \\ \diagup \\ R''\end{array} \longrightarrow \begin{array}{c}R' \\ \diagdown \\ C{-}OMgX \\ \diagup \\ R''\end{array} \longrightarrow \begin{array}{c}R \\ \diagdown \\ R'{-}COH \\ \diagup \\ R''\end{array}$$

Addition of alcohols to aldehydes gives *hemiacetals*, and reaction of these with a further molecule of alcohol gives *acetals*.

$$\begin{array}{c}H \\ \overset{\delta+}{RC} \diagup \\ \| \\ O^{\delta-} \\ \overset{\delta-}{EtO} \leftarrow \overset{\delta+}{H}\end{array} \longrightarrow \underset{\text{a hemiacetal}}{\begin{array}{c}H \\ \diagup \\ RC{-}OEt \\ \diagdown \\ OH\end{array}} \xrightarrow{EtOH} \underset{\text{an acetal}}{\begin{array}{c}H \\ \diagup \\ RC{-}OEt \\ \diagdown \\ OEt\end{array}}$$

The reaction occurs readily in the presence of dilute acid (e.g. HCl) but ketones do not readily give the analogous ketals under these conditions.

Several **compounds which can be thought of as H^+X^-** add to aldehydes and ketones thus:

$$\begin{matrix} \backslash^{\delta+}\ \ \ ^{\delta-} \\ C{=}O \\ / \\ X^-\ \ H^+ \end{matrix} \longrightarrow \begin{matrix} \backslash \\ C{-}OH \\ /| \\ X \end{matrix}$$

Thus *sodium hydrogensulphite* gives '*bisulphite compounds*' with aldehydes and ketones.

$$\rangle C{=}O + NaHSO_3 \longrightarrow \rangle C\langle^{OH}_{SO_3^-Na^+}$$

The mechanism is thought to involve the hydrogensulphite and sulphite ions as follows.

$$HSO_3^- \rightleftharpoons H^+ + SO_3^{2-}$$

$$\rangle \overset{\delta+}{C}{=}\overset{\delta-}{O} + SO_3^{2-} \longrightarrow \rangle C\langle^{O^-}_{SO_3^-} \xrightleftharpoons{HSO_3^-} \rangle C\langle^{OH}_{SO_3^-} + SO_3^{2-}$$

The bisulphite compounds are crystalline solids and readily isolated. They give back the aldehyde or ketone by hydrolysis on warming with dilute acid or alkali. This process therefore affords a method of purifying carbonyl compounds, or of separating them from mixtures with non-carbonyl compounds.

Hydrogen cyanide ($NaCN + H_2SO_4$) gives *cyanohydrins* with aldehydes and simpler ketones.

$$\rangle C{=}O + HCN \longrightarrow \rangle C\langle^{OH}_{CN}$$

Ammonia adds to aldehydes [other than methanal (formaldehyde)] but not to ketones, to give *aldehyde ammonias*. Methanal (formaldehyde) and ketones undergo more complex reactions with ammonia.

$$\begin{matrix} R \\ H \end{matrix}\rangle C{=}O + NH_3 \longrightarrow \begin{matrix} R \\ H \end{matrix}\rangle C\langle\begin{matrix} OH \\ NH_2 \end{matrix}$$

Sometimes addition to the carbonyl group is followed rapidly by loss of a molecule of water. Such **addition-elimination reactions** are undergone particularly with compounds of the form $H_2N{-}X$. The addition stages follow the same pattern as the simple additions just discussed and probably proceed by analogous mechanisms.

$$>C{=}O + H_2N{-}X \longrightarrow \left[>C\begin{smallmatrix} OH \\ NH{-}X \end{smallmatrix} \right] \xrightarrow{-H_2O} >C{=}NX$$

The nitrogen compounds themselves probably act as the nucleophiles in the addition, since like ammonia they have an unshared pair of electrons. A proton would then be transferred from nitrogen to oxygen in a subsequent stage.

$$\overset{\delta+}{>C}{=}\overset{\delta-}{O} \quad :NH_2X \longrightarrow >C\begin{smallmatrix} O^- \\ \overset{+}{N}H_2X \end{smallmatrix} \longrightarrow >C\begin{smallmatrix} OH \\ NHX \end{smallmatrix}$$

The intermediate addition compounds are not normally isolated. The complete processes have often been called '*condensation*' reactions, although the term 'addition-elimination' is preferable.

Thus *hydroxylamine* gives *oximes* (aldoximes and ketoximes):

$$>C{=}O + H_2N.OH \longrightarrow >C{=}NOH$$

Hydrazine gives *hydrazones*:

$$>C{=}O + H_2NNH_2 \longrightarrow >C{=}NNH_2$$

Arylhydrazines like phenylhydrazine, 4-(*p*-)-nitrophenylhydrazine, and 2,4-dinitrophenylhydrazine are more useful and give usually solid, crystalline *phenylhydrazones*, *4-*(p-)*-nitrophenlhydrazones*, and *2,4-dinitro-phenylhydrazones* respectively. The nitro-groups serve to raise the melting points of the derivatives.

$$>C{=}O + H_2NNHC_6H_5 \longrightarrow >C{=}NNHC_6H_5$$

phenylhydrazones

$$>C{=}O + H_2NNH{-}C_6H_4{-}NO_2 \longrightarrow >C{=}NNH{-}C_6H_4{-}NO_2$$

4-(*p*-)-nitrophenylhydrazones

$$>C{=}O + H_2NNH{-}C_6H_3(O_2N)(NO_2) \longrightarrow >C{=}NNH{-}C_6H_3(O_2N)(NO_2)$$

2,4-dinitrophenylhydrazones

These derivatives are easily isolated and purified by crystallization. They have characteristic melting points and are therefore used to identify the parent carbonyl compounds.

Semicarbazide similarly gives crystalline *semicarbazones*, which are also used for indentification.

$$\rangle C{=}O + H_2NNHCO.NH_2 \longrightarrow \rangle C{=}NNHCO.NH_2$$

semicarbazide — semicarbazones

Amines give products known as '*Schiff's bases*', e.g. with aniline (cf. Chapter 12, page 140).

$$\rangle C{=}O + H_2NC_6H_5 \longrightarrow \rangle C{=}NC_6H_5$$

a Schiff's base

Other carbonyl compounds can also act as addenda in addition reactions to aldehydes and ketones. The requirement is that a carbon atom adjacent to the carbonyl group (the α-carbon atom) must have at least one hydrogen attached to it. This is the hydrogen atom which becomes detached to leave a carbanion, which is the nucleophile. This process is assisted because the carbonyl group is strongly attracting (remember that it is its positive end which is nearer to the rest of the molecule). The process is strongly assisted by basic catalysts (e.g. EtO^- from sodium ethoxide) and such catalysts are necessary to bring about these reactions.

$$R_2C(H){-}C{=}O \xrightarrow{EtO^-} [R_2\bar{C}{-}C{=}O \longleftrightarrow R_2C{=}C{-}O^-] + EtOH$$

The carbanion is stabilized by the resonance shown.

This addition reaction is known generally as the *aldol reaction*, and there are many variations on it. Although it is in principle reversible, the equilibrium often lies well to the right. The simplest example, given below, is that in which ethanal (acetaldehyde) acts as both addendum and acceptor. The product is aldol itself, and the catalyst is usually dilute aqueous sodium carbonate.

$$CH_3C{=}O \;\; OHC.CH_2{-}H \underset{Na_2CO_3}{\overset{dil.}{\rightleftharpoons}} CH_3C(H)({-}OH)CH_2CHO \quad [CH_3C(OH)CH_2CHO]$$

aldol
(3-hydroxybutanal)

The reaction can be followed by loss of water, because the addition products are readily dehydrated to give unsaturated carbonyl compounds, e.g.

$$CH_3CH(OH)CH_2CHO \longrightarrow \underset{\substack{\text{crotonaldehyde}\\ \text{(but-2-enal)}}}{CH_3CH{=}CHCHO}$$

These products are themselves carbonyl compounds so further addition to them can take place. This occurs extensively if aliphatic aldehydes are heated with caustic alkali, when brown resins of high relative molecular mass are formed by successive additions and dehydrations.

Aromatic aldehydes like benzaldehyde are good acceptors in the aldol-type reactions because of the anyl group (cf. page 105). They cannot act as addenda, however, because the carbon atom adjacent to the carbonyl group (the α-carbon atom) has no hydrogen attached to it. Nor, for the same reason, do they give resins with caustic alkali. Instead they undergo a different process which is known as the *Cannizzaro reaction.* This is a reaction in which one molecule of benzaldehyde (for example) is oxidized to benzoic acid at the expense of another, which is reduced to phenylmethanol (benzyl alcohol).

$$2C_6H_5CHO \xrightarrow{NaOH} \underset{\substack{\text{phenylmethanol}\\ \text{(benzyl alcohol)}}}{C_6H_5CH_2OH} + \underset{\text{benzoic acid}}{C_6H_5CO_2H}$$

The Cannizzaro reaction is undergone by methanal (formaldehyde) but not by other aliphatic aldehydes, which, having α-carbon atoms to which hydrogen is attached, can undergo aldol reactions.

$$2H_2CO \xrightarrow{NaOH} \underset{\text{methanol}}{CH_3OH} + \underset{\substack{\text{formic acid}\\ \text{(methanoic acid)}}}{HCO_2H}$$

Polymerization

Reactions in which two or more molecules of the same substance react together to give a more complex molecule and no other products are called *polymerization* (Chapter 6, page 51). Thus the formation of aldol from ethanal is a polymerization (or, more specifically, dimerization). Polymerizations are thus essentially addition reactions.

Ethanal polymerizes in other ways, also. With concentrated sulphuric acid, it gives the liquid *trimer 'paraldehyde'*.

$$3CH_3CHO \underset{\substack{\text{heat}\\ \text{dil. } H_2SO_4}}{\overset{\text{conc. } H_2SO_4}{\rightleftharpoons}} \begin{array}{c} \quad O \quad \\ CH_3CH \qquad CHCH_3 \\ | \qquad\quad | \\ O \qquad\quad O \\ CH \\ | \\ CH_3 \end{array}$$

ethanal trimer
(paraldehyde)

With acids below 0 °C it gives the crystalline *ethanal tetramer 'metaldehyde'*

$[(CH_3CHO)_4]$, which probably has an analogous structure. Metaldehyde is sometimes used as a convenient solid fuel (e.g. for picnic stoves).

Methanal (formaldehyde) also gives an analogous crystalline trimer '*trioxan*' on boiling its aqueous solution with sulphuric acid. Methanal can be regenerated by heating the trimer.

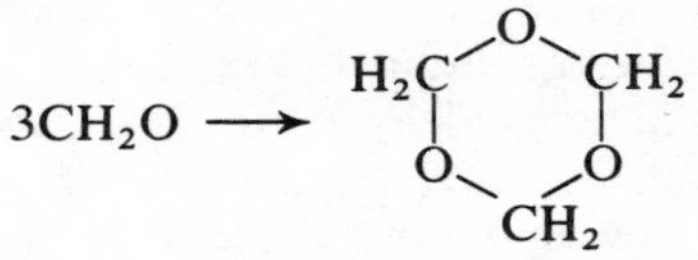

$$3CH_2O \longrightarrow \text{(cyclic } (CH_2O)_3\text{: ring of } H_2C, O, CH_2, O, CH_2, O)$$

trioxan

With lime water methanal gives first the dimer which is hydroxyethanal, and then a mixture of *monosaccharide sugars* (cf. Chapter 13, page 151) called 'formose'. These monosaccharides can be regarded as hexamers of methanal, i.e. $(CH_2O)_6$.

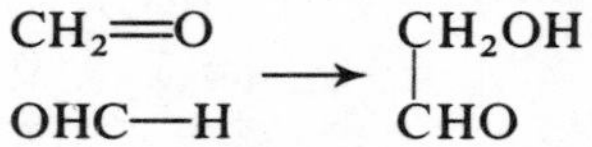

$$\begin{matrix} CH_2{=}O \\ OHC{-}H \end{matrix} \longrightarrow \begin{matrix} CH_2OH \\ | \\ CHO \end{matrix}$$

hydroxyethanal

$$6CH_2O \longrightarrow C_6H_{12}O_6$$

'formose'

For this reason formaldehyde was once suggested as an intermediate in photosynthesis, which is the natural process by which plants synthesize carbohydrates (of which group of compounds the monosaccharides are simple members, cf. Chapter 13) from carbon dioxide and water.

Halogenation

Carbonyl compounds react with halogenating agents like phosphorus pentachloride to give gem-dihalides:

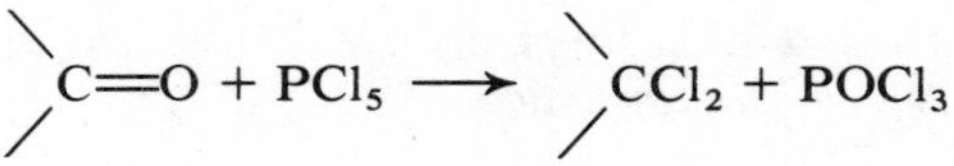

$$\gt C{=}O + PCl_5 \longrightarrow \gt CCl_2 + POCl_3$$

This reaction may be rationalized by considering it as proceeding through the hypothetical hydrate of the carbonyl compound, which being a gem-dihydroxy-compound would be expected to be highly susceptible to loss of water so that the equilibrium:

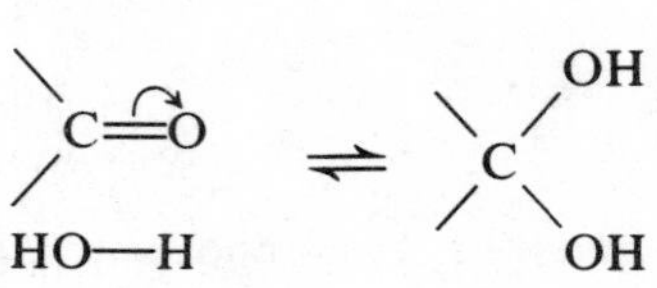

$$\begin{matrix} \gt C{=}O \\ HO{-}H \end{matrix} \rightleftharpoons \gt C \begin{matrix} OH \\ OH \end{matrix}$$

usually lies overwhelmingly to the left. However, in the presence of a halogenating agent, the hydroxyl groups in the gem-dihydroxy-compound are replaced by chlorine. The above equilibrium is thus continuously disturbed, more dihydroxy-compound is formed, and the reaction can proceed ultimately to completion.

A very few stable hydrates do exist, e.g. 'chloral hydrate':

$$\underset{\text{trichloroethanal ('chloral')}}{CCl_3CHO} + H_2O \longrightarrow \underset{\text{2,2,2-trichloroethanediol ('chloral hydrate')}}{CCl_3CH(OH)_2}$$

Chloral has been used as a hypnotic.

The Haloform reaction

This reaction is undergone by methyl ketones (i.e. ketones in which at least one of the alkyl groups is methyl). The reagent is the sodium oxohalate(I) (hypohalite) and the products are the haloform (the trihalogenomethane) and the salt of a carboxylic acid, containing one fewer carbon atom than the ketone. It is used for the preparation of *either* of these products, e.g. with sodium oxochlorate(I) (hypochlorite):

$$R{-}\underset{\underset{O}{\|}}{C}{-}CH_3 + 3NaOCl \longrightarrow R{-}\underset{\underset{O}{\|}}{C}{\mid}{-}CCl_3\ (+\ 3NaOH) \xrightarrow[(OH^-)]{\text{hydrolysis}}$$

$$\underset{\text{anion of carboxylic acid}}{R{-}C\begin{matrix}\nearrow O \\ \searrow O^-\end{matrix}} + \underset{\text{trichloromethane (chloroform)}}{CHCl_3}$$

Similarly iodine and sodium hydroxide (which can be regarded as containing the oxoiodate(I) ion) with, for example, acetone gives triiodomethane (iodoform):

$$CH_3CO.CH_3 \xrightarrow{I_2/NaOH} CH_3CO.CI_3 \longrightarrow \underset{\text{acetate ion (ethanoate ion)}}{CH_3CO.O^-} + \underset{\text{triiodomethane (iodoform)}}{CHI_3}$$

The formation of the pale yellow crystalline, characteristically smelling, triiodomethane (iodoform) is a test for a methyl ketone.

The haloform reaction is also undergone by ethanal (not by other aldehydes, since the *methyl* group is essential). Ethanol is, however, more commonly used as the starting material in this case because it is oxidized *in situ* to ethanal, e.g.

$$CH_3CH_2OH \xrightarrow[\text{NaOCl or bleaching powder}]{(O)} CH_3CHO \xrightarrow{3ClO^-} CCl_3CHO \ (+\ 3NaOH)$$

$$CCl_3CHO \xrightarrow{\text{hydrolysis } (OH^-)} CHCl_3 + HCO.O^-$$

formate ion
(methanoate ion)

This is how trichloromethane is usually prepared. Triiodomethane can be obtained similarly by warming ethanol with iodine and sodium hydroxide solution.

Propan-2-ol also undergoes the haloform reaction because it is first oxidized to acetone, which reacts as above, e.g.

$$CH_3CH(OH)CH_3 \xrightarrow{I_2/NaOH} [CH_3CO.CH_3] \longrightarrow CH_3CO.O^- + CHI_3$$

Questions

1. (*a*) Give the reagent(s), equation(s) and *essential* reaction conditions for the conversion of ethyl alcohol (ethanol) to each of the following compounds: (i) ethyl bromide (bromoethane); (ii) diethyl ether (ethoxyethane); (iii) acetaldehyde (ethanal); (iv) ethylene (ethene); (v) acetic acid (ethanoic acid).
   (*b*) Give the equation, and name the product formed, when acetaldehyde (ethanal) reacts with each of the following: (i) sodium bisulphite; (ii) 2,4-dinitrophenylhydrazine; (iii) iodine and alkali.
   (*c*) Give the reaction which occurs when acetaldehyde (ethanal) reacts with each of the following (equations are **not** required): (i) ammoniacal silver nitrate; (ii) Fehling's solution.
   (WJEC, 1974)
2. Explain the reactions shown in the following scheme and assign structures to the substances **A** to **H** inclusive.

$$\underset{\textbf{A}}{C_2H_4O} \xrightarrow{\text{dil. NaOH(aq)}} CH_3CH(OH)CH_2CHO \xrightarrow{Ag_2O} \underset{\textbf{B}}{C_4H_8O_3}$$

$$CH_3CH(OH)CH_2CHO \xrightarrow{NaBH_4} \underset{\textbf{C}}{C_4H_{10}O_2}; \quad \xrightarrow{\text{heat}} \underset{\textbf{D}}{C_4H_6O}; \quad \xrightarrow{HCN} \underset{\textbf{E}}{C_5H_9NO_2}$$

$$\underset{\textbf{D}}{C_4H_6O} \xrightarrow{Br_2} \underset{\textbf{F}}{C_4H_6Br_2O}; \quad \xrightarrow{Ag_2O} \underset{\textbf{G}}{C_4H_6O_2}; \quad \xrightarrow{NH_2OH} \underset{\textbf{H}}{C_4H_7NO}$$

   (O and C, 1973)
3. What products are formed when (*a*) propanone (acetone) is treated with hydroxylamine and (*b*) ethyl ethanoate (acetate) is boiled with water? What catalysts may be used to increase the rate of reaction (*b*) and how do these catalysts function?

What products are formed when the compound $(CH_3)_2C(OH)CN$, obtained from reaction between propanone and hydrogen cyanide, reacts with (i) dilute aqueous sodium hydroxide, (ii) aqueous hydrochloric acid?

Treatment of propanone with chlorine and aqueous sodium hydroxide yields trichloromethane (chloroform). What mass of trichloromethane may be obtained from 1.00 g of propanone?

[Relative atomic masses are H 1, C 12, O 16, Cl 35.5.]

(O and C, 1975)

4. Give an example of an *aldehyde*, a *ketone* and a *carboxylic acid*, in each case showing fully the structure of the characteristic functional group.

Describe, without experimental details, one method by means of which benzaldehyde may be obtained from toluene.

Give **two** reactions, each of which is common to benzaldehyde and acetaldehyde, and point out **one** way in which the chemical properties of these two compounds differ. Explain briefly the difference that you mention.

(O and C, 1972)

5. How would you obtain acetone (propan-2-one) from (*a*) propylene (propene) (*b*) acetic acid (ethanoic acid)?

What products are obtained when acetone is reacted with (i) hydroxylamine (ii) sodium hypochlorite solution (iii) deuterium oxide and a small quantity of sodium hydroxide (iv) lithium aluminium hydride (v) sodium hydrogen sulphite solution?

(O and C, 1971)

6. What is meant by a carbonyl group? Give examples of common types of organic compounds which contain such a group.

Discuss the reactions of these compounds containing carbonyl groups, including in your answer a brief comparison with the reactions of alkenes. Reaction mechanisms should be given where possible.

(JMB, 1975)

7. Aldehydes, ketones and carboxylic acids contain the carbonyl group ($>C{=}O$). This group is often said to be responsible for the addition and condensation reactions undergone by aldehydes and ketones.

(*a*) Review briefly the major addition and condensation reactions shown by aldehydes and ketones; in each case discussed, examine the role of the carbonyl group in bringing about these reactions.

(*b*) Carboxylic acids do not undergo any of the addition and condensation reactions shown by aldehydes and ketones, despite the fact that they contain the carbonyl group. Give explanations for this.

(London, 1972)

Chapter 11

Carboxyl and Related Functions: Stereochemistry

Aliphatic, 'fatty', or carboxylic acids contain the carboxyl group, which itself contains a *carb*onyl and a hydr*oxyl* group.

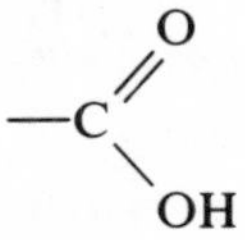

$$-C\overset{\displaystyle O}{\underset{\displaystyle OH}{\lessgtr}}$$

Monocarboxylic acids are monobasic, since only the carboxyl group hydrogen atom is replaceable by metals. Dicarboxylic acids are dibasic, etc. The carboxyl group can be attached to saturated, unsaturated, or aromatic frameworks.

The carboxyl group should be written as $-CO_2H$ or as $-CO.OH$, with the full stop after the carbonyl group. It should *never* be written $-COOH$ because of the possible confusion with hydroperoxides, which are correctly written ROOH, since the two oxygen atoms are linked together in them, i.e. R–O–O–H.

Nomenclature

The simpler members are usually known by their trivial names. The systematic group suffix is -oic acid, as in the following examples.

HCO_2H
methanoic acid
(formic acid)
(m.p. 8.4 °C; b.p. 100.5 °C)

CH_3CO_2H
ethanoic acid
(acetic acid)
(m.p. 16.6 °C; b.p. 118 °C)

$CH_3CH_2CO_2H$
propanoic acid
(propionic acid)
(m.p. −22 °C; b.p. 141 °C)

$CH_3CH_2CH_2CO_2H$
butanoic acid
(n-butyric acid)
(m.p. −4.7 °C; b.p. 162 °C)

$(CH_3)_2CHCO_2H$
2-methylpropanoic acid
(isobutyric acid)
(m.p. −4.7 °C; b.p. 154 °C)

The C_5 acid, pentanoic acid, is called valeric acid, and the C_6 acid, hexanoic acid, is called caproic acid.

Acids containing more than one carboxyl group are called -dioic, -trioic, etc. acids. Thus, for example, the acid $HO_2CCH_2CH_2CH_2CH_2CO_2H$, which is commonly known as adipic acid, is systematically named hexanedioic acid. There is no need for numbering, since the carboxyl groups can only be at the ends of the chain.

The aromatic acid $C_6H_5CO_2H$ is benzoic acid (or benzenecarboxylic acid). It is sometimes convenient to regard the carboxyl group as a substituent, e.g.

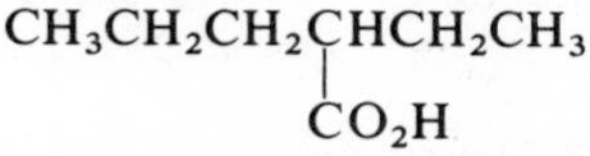

$$CH_3CH_2CH_2\underset{\displaystyle CO_2H}{\underset{|}{C}}HCH_2CH_3$$

hexane-3-carboxylic acid
(or 2-ethylpentanoic acid)

The groups $R{-}C(=O){-}$ ($RCO{-}$) are known as *acyl* groups, e.g.

$HC(=O){-}$ formyl (methanoyl), $CH_3C(=O){-}$ acetyl (ethanoyl), etc.

Formation of the carboxyl group

Methods are merely mentioned here, with appropriate cross references if detailed consideration is given elsewhere.

Oxidation of primary alcohols or aldehydes gives carboxylic acids (Chapter 9, page 92; Chapter 10, page 105). Thus vinegar, which contains acetic acid, is formed by the fermentation of wine, beer, or cider in the presence of bacteria (*mycoderma aceti*). The reaction is aerial oxidation.

$$CH_3CH_2OH + O_2 \longrightarrow CH_3CO_2H + H_2O$$

Manganese(II) salts also catalyse the aerial oxidation of ethanal to acetic acid. Aromatic side chains can be oxidized quite readily to carboxyl groups. Many common oxidizing agents (e.g. alkaline permanganate) can accomplish this.

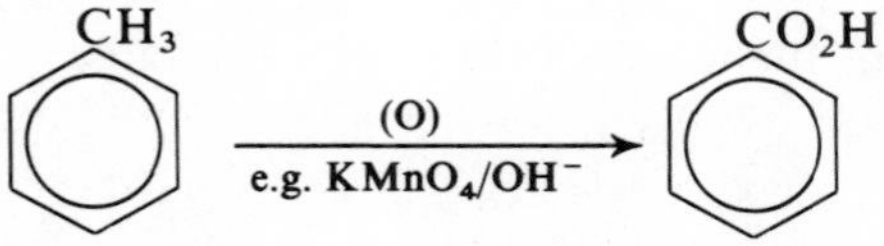

$$C_6H_5CH_3 \xrightarrow[\text{e.g. } KMnO_4/OH^-]{(O)} C_6H_5CO_2H$$

methylbenzene (toluene) → benzoic acid

Grignard reagents react with carbon dioxide to give carboxylic acids. The reaction is simply the addition of the Grignard reagent to a carbonyl group, as previously described (Chapter 10, page 107).

$$RMgX + O{=}C{=}O \longrightarrow O{=}\overset{R}{\overset{|}{C}}{-}OMgX \xrightarrow{\text{hydrolysis}} R{-}C(=O){-}OH$$

This affords a method of extending the carbon chain of an alcohol (for example) by one carbon atom, by the following series of reactions.

$$ROH \xrightarrow{Br_2/\text{red P}} RBr \xrightarrow{\text{Mg/dry ether}} RMgBr \xrightarrow{CO_2}$$

$$RCO_2H \xrightarrow[\text{(LiAlH}_4\text{, cf. page 123)}]{\text{reduction}} RCH_2OH$$

There are also other methods of ascending (and descending) a homologous series, and these will be described at appropriate points (cf. Chapter 12, pages 141, 148).

Benzoic acid can be synthesized from benzene by the following reactions.

$$C_6H_6 \xrightarrow{Br_2} C_6H_5Br \xrightarrow{\text{Mg/dry ether}} C_6H_5MgBr \xrightarrow{CO_2} C_6H_5CO_2H$$

Hydrolysis of nitriles (cyanides) gives carboxylic acids as their ammonium salts. (Chapter 12, page 140.)

$$RCN \xrightarrow{2H_2O} RCO_2NH_4$$

Carboxylic acids are also formed by hydrolysis of esters (page 125), acyl halides (page 127), amides (page 129), and anhydrides (page 128).

Other methods, which are frequently the most appropriate in particular cases, are available, but are beyond the scope of this book.

Properties and reactions of carboxylic acids

The lower members are liquids with pungent smells and are soluble in water. The C_4—C_9 acids are oily substances, sparingly soluble in water, with unpleasant smells. The higher acids are solids, insoluble in water. Benzoic acid is a solid, m.p. 121 °C, soluble in hot water, but only very sparingly soluble in cold water.

Acidity

The most important property of the carboxylic acids is that they are weak acids. The acidity arises from the presence of the strongly electron-attracting carbonyl group, which facilitates the separation of the hydroxyl hydrogen as a proton in a polar solvent such as water:

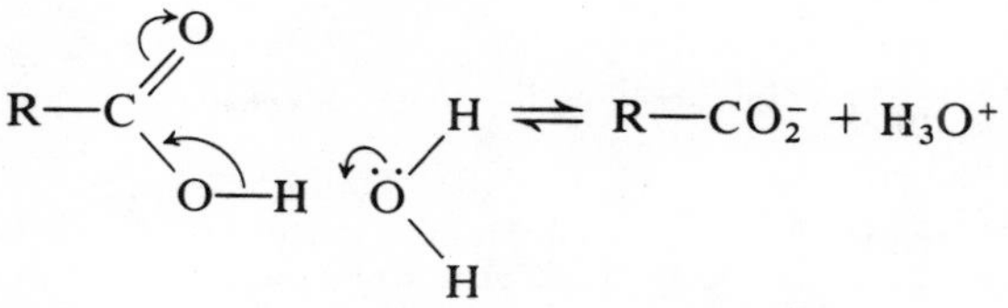

$$R{-}C(=O){-}O{-}H + H_2O \rightleftharpoons R{-}CO_2^- + H_3O^+$$

Another way of looking at this is that the *carboxylate anion* is greatly stabilized by the resonance:

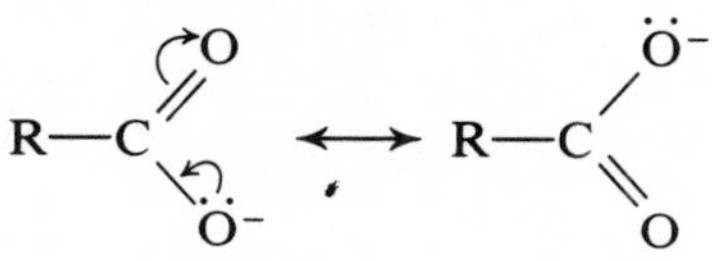

$$R{-}C(=O){-}\ddot{O}^- \longleftrightarrow R{-}C(\ddot{O}^-){=}O$$

which arises from the same cause. The double bond is not localized *in the carboxylate anion*, but occupies a molecular orbital covering both oxygens and the carbon. The ion might therefore be more accurately represented thus:

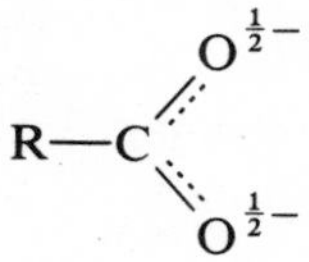

$$R{-}C(\cdots O^{\frac{1}{2}-})(\cdots O^{\frac{1}{2}-})$$

For this reason the acidity of the hydroxyl group is very much enhanced by the presence of the carbonyl group adjacent to it. An extension of this concept is that the acid strength is increased by the presence of electron attracting substituents in R, and decreased by the presence of electron-repelling substituents. This is illustrated by the following pK_a values (remember that the higher the pK_a, the weaker the acid).

	$H{-}CO_2H$	$CH_3{\rightarrow}CO_2H$	$CH_3{\rightarrow}CH_2{\rightarrow}CO_2H$
	formic acid (methanoic acid)	acetic acid (ethanoic acid)	propanoic acid
pK_a	3.75	4.76	4.87

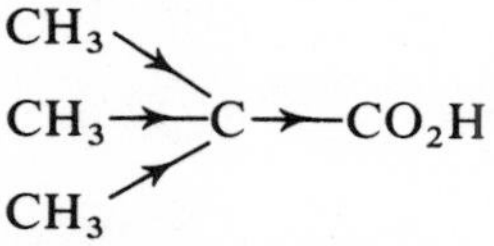

$$(CH_3)_3{\rightarrow}C{\rightarrow}CO_2H$$

2,2-dimethylpropanoic acid
(trimethylacetic acid)

pK_a 5.05

The effect is very clearly illustrated with the halogenoacetic acids, which are progressively strengthened by the electron attraction of the electronegative halogen atoms:

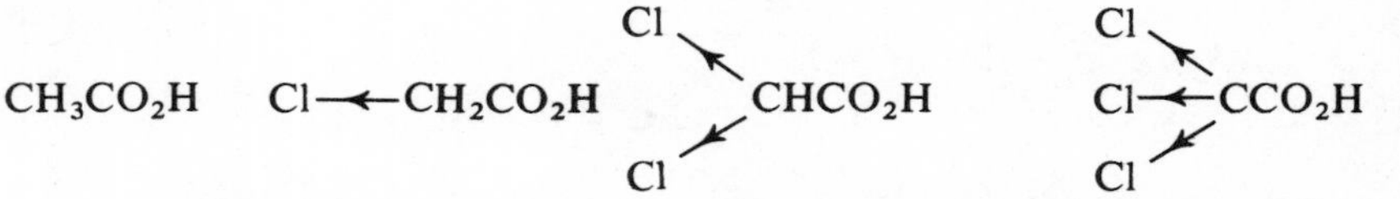

CH_3CO_2H	$Cl \leftarrow CH_2CO_2H$	$Cl_2 \leftarrow CHCO_2H$	$Cl_3 \leftarrow CCO_2H$
acetic acid (ethanoic acid)	chloroacetic acid (chloroethanoic acid)	dichloroacetic acid (dichloroethanoic acid)	trichloroacetic acid (trichloroethanoic acid)
pK_a 4.76	2.85	1.25	0.66

The strengths of other acids can be interpreted on the basis of similar concepts. It must however be remembered that these are all weak acids (cf. HCl has a pK_a of about -7).

Lack of carbonyl properties

The second striking feature about this group is that neither the acids nor their salts display the characteristic reactions of the carbonyl group (Chapter 10). In the salts (anions) the explanation is obvious: the resonance of the carboxylate anion group means that no carbonyl group as such is present.

In the free acids, this highly developed resonance is not possible but still carbonyl reactions are absent. No doubt the polarization:

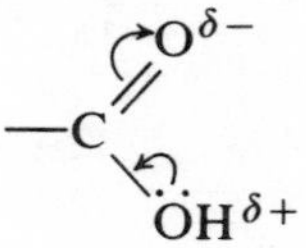

$$-C(=O^{\delta-})-\ddot{O}H^{\delta+}$$

is part of the explanation, since the supply of electron density by the hydroxyl oxygen means that it, rather than the carbonyl carbon, develops the slight positive charge. The reactivity of the carbonyl carbon towards nucleophiles is therefore not enhanced as it would be if the hydroxyl group were not present.

There is also, however, another important contributory factor. In the liquid phase, and in solution, carboxylic acids are associated by hydrogen-bonding, forming cyclic dimers. Thus if relative molecular masses of these acids are measured by solution methods, double values are obtained (e.g. acetic acid in benzene has r.m.m. 120). The structure of these dimers is:

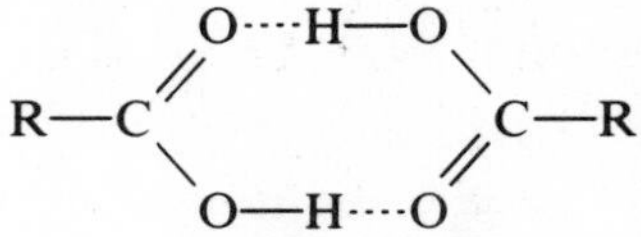

$$R-C\begin{matrix} \diagup\!\!\diagup O\cdots H-O \diagdown \\ \diagdown O-H\cdots O \diagup\!\!\diagup \end{matrix}C-R$$

Before carbonyl reactions can occur the hydrogen bond must be broken, and since this itself requires about 25 kJ mol^{-1}, these reactions are effectively suppressed.

Halogenation

Halogenation of the alkyl groups takes place with chlorine or bromine in the presence of light or a catalyst (e.g. iodine or red phosphorus). The α-position is the most reactive, e.g.

$$CH_3CO_2H \xrightarrow[h\nu]{Cl_2} CH_2Cl.CO_2H \longrightarrow CHCl_2.CO_2H \longrightarrow CCl_3.CO_2H$$

These are usually homolytic reactions with mechanisms analogous to the halogenation of alkanes (Chapter 5, page 40).

The hydroxyl group can be replaced by halogens by the use of halogenating agents (cf. Chapter 9, page 92), e.g.

$$RCO.OH + PCl_5 \longrightarrow RCO.Cl + HCl + POCl_3$$
$$RCO.OH + SOCl_2 \longrightarrow RCO.Cl + HCl + SO_2$$

The products are *acid chlorides* (or *acyl* chlorides; cf. page 127).

Dehydration

Anhydrides are formed by loss of a molecule of water from two molecules of a carboxylic acid, using for example phosphorus(V) oxide as dehydrating agent.

$$2CH_3CO.OH \xrightarrow[-H_2O]{P_2O_5} CH_3CO.O.CO.CH_3$$

acetic anhydride
(ethanoic anhydride)

Esterification and Ester Hydrolysis

The esterification reaction is reversible, so it is convenient to treat esterification and ester hydrolysis together.

acid + alcohol ⇌ ester + water

e.g.

$$CH_3C(=O)OH + C_2H_5OH \rightleftharpoons CH_3C(=O)OC_2H_5 + H_2O$$

ethyl acetate
(ethyl ethanoate)

Esterification must be carried out in acid conditions, since in alkaline solution the anion $RCO.O^-$ is formed, and this will not react. The equilibrium is usually disturbed by distilling out the ester as it is formed, so that the reaction goes to completion. The acid catalyst is present as a proton-donor.

Ester hydrolysis can take place by different mechanisms under acid or alkaline conditions. Alkaline hydrolysis is usually preferred since the product is the carboxylate anion, which does not undergo esterification, so the hydrolysis proceeds to completion.

Detailed consideration of the mechanisms of esterification and ester hydrolysis is beyond the scope of this book, because of the multiplicity of possible mechanisms, most of which have actually been observed under various conditions. There are in fact four possible mechanisms for esterification and eight for hydrolysis. This arises because (1) esterification must be acid catalysed whereas hydrolysis can be either acid or base catalysed, so that only half of the mechanisms are available for esterification; (2) there are unimolecular and bimolecular variations of each fundamental mechanistic scheme; and (3) there are two possible bonds which may be broken. Thus for hydrolysis, position (a) is called 'acyl oxygen fission' and position (b) 'alkyl oxygen fission'.

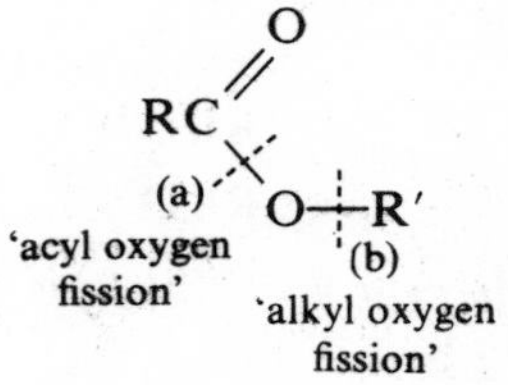

RC(=O)–O–R′ with bond (a) between RC and O marked 'acyl oxygen fission' and bond (b) between O and R′ marked 'alkyl oxygen fission'

The position of fission in any particular reaction can be determined by the use of water enriched with the ^{18}O isotope. The following results are then obtained (for hydrolysis).

Acyl oxygen fission:

$$RCO.\vdots OR' + H^{18}OH \rightleftharpoons RCO.^{18}OH + HOR'$$

(i.e. label in the *acid*)

Alkyl oxygen fission:

$$RCO.O\vdots R' + H^{18}OH \rightleftharpoons RCO.OH + H^{18}OR'$$

(i.e. label in the *alcohol*)

The position of the label in the product thus indicates the position of fission. It is obviously easier to perform this test for hydrolysis than for esterification. However, the conclusion reached is also applicable to the reverse esterification under identical conditions. This is because the mechanisms of forward and reverse reactions taking place together in a reversible reaction must be identical, i.e. the transition state(s) must be common to both forward and reverse reactions. This is important enough to be given a special name: it is called the *Principle of Microscopic Reversibility*. Ingold expressed it as equivalent to the statement that the easiest route from Borrowdale to Wasdale (two valleys in the English Lake District) is the same as the easiest route from Wasdale to Borrowdale. The importance of the Principle is that one can study either the forward or the reverse reaction, whichever is more convenient, and apply the conclusions to the other.

It is appropriate to give here the mechanisms of hydrolysis and esterification under the most common conditions when, as it turns out, acyl oxygen fission usually occurs. Alkaline ester hydrolysis is usually bimolecular and the mechanism is:

$$R{-}C(OR'){=}O + {}^{-}\!:\!OH \underset{slow}{\rightleftharpoons} R{-}C(OH)(OR'){-}O^{-} \xrightarrow{fast} R{-}C(OH){=}O + {}^{-}OR' \xrightarrow{fast} R{-}C(O^{-}){=}O + R'OH$$

It is worth noting that the reaction involves full addition of the hydroxide ion to the carboxyl carbon to give an *intermediate* rather than synchronous approach of $^{-}$OH and recession of OR′ (as in the S_N2 mechanism), which would mean the reaction proceeding through a *transition state*. This becomes possible because the carbonyl oxygen is now available to accommodate the negative charge in the intermediate.

The most common mechanism for esterification (and therefore by the principle of microscopic reversibility, also for acid catalysed hydrolysis) is also bimolecular and involves acyl oxygen fission. It must therefore be the *acid* which is protonated by the acid catalyst.

$$R{-}C(OH){=}O + H^{+} \underset{}{\overset{fast}{\rightleftharpoons}} R{-}C(\overset{+}{O}H_2){=}O \quad \left(\overset{fast}{\rightleftharpoons} R{-}\overset{+}{C}(OH){-}OH \right)$$

$$+\ H{-}\ddot{O}{-}R'$$

$$\text{slow} \downarrow\uparrow \text{fast}$$

$$R{-}C(\overset{+}{O}H_2)(\overset{+}{O}HR'){-}O^{-} \quad \left(\overset{fast}{\rightleftharpoons} R{-}C(OH)(OH){-}\overset{+}{O}HR' \right)$$

$$\text{fast} \downarrow\uparrow \text{slow}$$

$$H_2O: \quad R{-}C(\overset{+}{O}HR'){=}O \overset{fast}{\rightleftharpoons} R{-}C({=}O){-}OR' + H^{+}$$

The protonated intermediates can be written in two ways because of the possibility of proton transfer between the hydroxyl and carbonyl oxygens.

Although these are the most common mechanisms, it must be emphasized that others, involving alkyl oxygen fission, unimolecular processes, and protonation of the alcohol instead of the acid, can operate under appropriate conditions.

Reduction

Carboxylic acids can be reduced to primary alcohols, but only with the vigorous reducing agent lithium(I)tetrahydridoaluminate(III) (lithium aluminium hydride).

$$RCO_2H \xrightarrow{LiAlH_4} RCH_2OH$$

Because of the vigour of the reagent, the reaction cannot be stopped at the intermediate aldehyde stage (Chapter 10, page 106).

Formic acid (methanoic acid)

Formic acid is unique because it contains an aldehyde group as well as a carboxyl group.

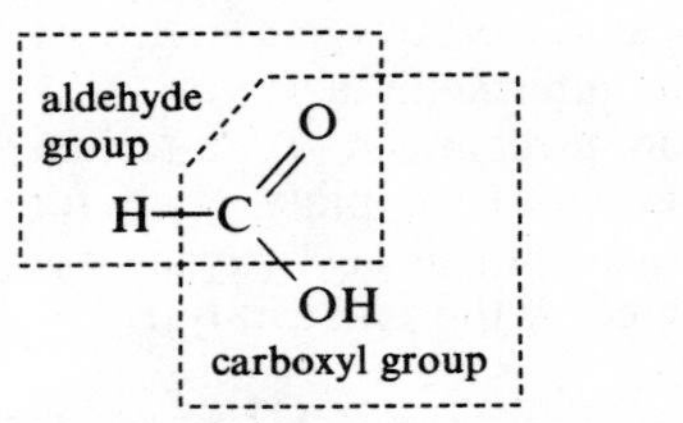

It is therefore readily oxidized to carbon dioxide and water (carbonic acid). It is a reducing agent, and gives positive tests with Fehling's solution and Tollen's reagent.

$$HCOOH + (O) \longrightarrow \left[(HO)_2C{=}O \right] \longrightarrow H_2O + CO_2$$

carbonic acid

It can be prepared by oxidation of methanol or methanal (formaldehyde).

$$CH_3OH \xrightarrow[(-H_2O)]{(O)} HCHO \xrightarrow{(O)} HCO_2H$$

It is also obtained by *decarboxylation* of (loss of carbon dioxide from) ethanedioic (oxalic) acid. This is usually done by heating oxalic acid with glycerol (propane-1,2,3-triol).

$$\begin{matrix} CO_2H \\ | \\ CO_2H \end{matrix} \xrightarrow{-CO_2} \begin{matrix} H \\ | \\ CO_2H \end{matrix}$$

ethanedioic acid (oxalic acid) — formic acid

It is obtained industrially by heating together carbon monoxide and sodium hydroxide solution under pressure.

$$CO + NaOH \xrightarrow[6\text{–}10\ \text{atm.}]{\sim 200\ ^\circ C} HCO_2Na \xrightarrow{H_2SO_4} HCO_2H$$

Formic acid occurs naturally in nettles and in ants (hence its name). It can be dehydrated with concentrated sulphuric acid to give carbon monoxide.

$$HCO_2H \xrightarrow[(-H_2O)]{H_2SO_4} CO$$

Unlike the other members of the series it does not give an acid chloride. It is about twelve times stronger as an acid than acetic acid (cf. page 119).

Derivatives of carboxylic acids

Esters

Esters can be prepared by the esterification reaction, by reaction of alcohols with acid halides, or anhydrides (page 128), or by reaction of the silver salt of the acid with a haloalkane.

$$RCO_2Ag + R'I \longrightarrow RCO.OR' + AgI$$

The simpler esters are volatile liquids, insoluble in water, with pleasant 'fruity' smells. They occur in nature as the fragrant constituents of fruits, and as fats and waxes.

Esters can be hydrolysed as described above. They react with ammonia to give acid amides (cf. page 128).

$$RCO.OR' + HNH_2 \longrightarrow \underset{\text{amide}}{RCO.NH_2} + R'OH$$

An ester can be reduced to the primary alcohol corresponding to the acid from which it is derived by lithium(I) tetrahydridoaluminate(III) (lithium aluminium hydride), or by sodium and ethanol.

$$RCO.OR' \xrightarrow[\text{or Na/EtOH}]{LiAlH_4} RCH_2OH\,(+\,R'OH)$$

The reduction of the ester is easier than that of the acid itself (cf. page 123).

Esters show 'carbonyl' reactions to a small degree. Thus they react with Grignard reagents.

$$\overset{\delta+}{R}C(=O^{\delta-})OR' + R''^{-}MgX^{+} \longrightarrow (R)(R'')C(OMgX)(OR')$$

$$\xrightarrow{\text{dil. HCl}} (R)(R'')C{=}O \xrightarrow{R''MgX} R''{-}C(R)(R''){-}OMgX$$

$$(+\,R'OH + MgXCl) \xrightarrow{\text{dil. HCl}} R''{-}C(R)(R'')OH + MgXCl$$

The ketone so formed can react further with any excess of Grignard reagent present to give a tertiary alcohol in which two of the three alkyl groups attached to the α-carbon atom are the same, and this is often a convenient way of preparing such alcohols.

The reactivity of the carbonyl group in esters is reduced: they do not react with hydrogensulphite ions, hydrogen cyanide, hydroxylamine, phenylhydrazine, etc. This is due to the modifying effect of the polarization and resonance:

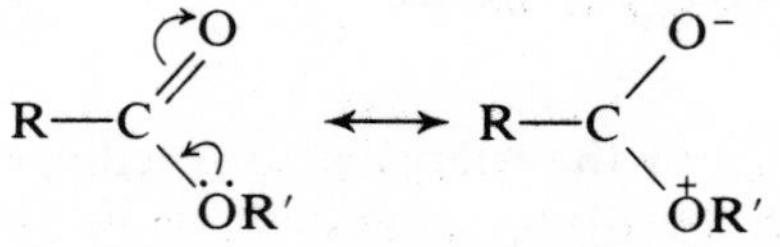

$$R{-}C(=O){-}\ddot{O}R' \longleftrightarrow R{-}C(-O^-){=}\overset{+}{O}R'$$

as in the acids themselves, although the further reduction in reactivity arising from dimerization of the acids is not observed in the esters, which consequently are somewhat more reactive as carbonyl compounds than the acids.

Oils and fats

Oils and fats are esters of long-chain acids with glycerol (propane-1,2,3-triol). They are derived from acids such as dodecanoic (lauric) acid (C_{12}), hexadecanoic (palmitic) acid (C_{16}), and octadecanoic (stearic) acid (C_{18}). The acids from which oils are derived are often unsaturated, like *cis*-octadec-9-enoic (oleic) acid (C_{18})*. These can be *hardened* by catalytic hydrogenation (e.g. Sabatier–Senderens, Ni catalyst, cf. Chapter 6, page 47) to give solid fats. This is the essential process in the manufacture of margarine.

Hydrolysis of fats gives glycerol together with the sodium salts of the acids, which are soaps, e.g.

$$\begin{matrix} CH_2O.CO.C_{17}H_{33} \\ | \\ CH.O.CO.C_{17}H_{33} \\ | \\ CH_2O.CO.C_{17}H_{33} \end{matrix} \xrightarrow{NaOH} \begin{matrix} CH_2OH \\ | \\ CHOH \\ | \\ CH_2OH \end{matrix} + 3C_{17}H_{33}CO_2Na$$

glyceryl tristearate 'tristearin'   glycerol

The detergent action of soaps (and indeed of other detergents) arises from the fact that the hydrocarbon end of the chain is soluble in organic solvents, and the salt end is soluble in water. The soap anion can thus form a 'bridge' between the greasy dirt particle and the washing water, thus enabling the dirt to be wetted and therefore washed away by the water:

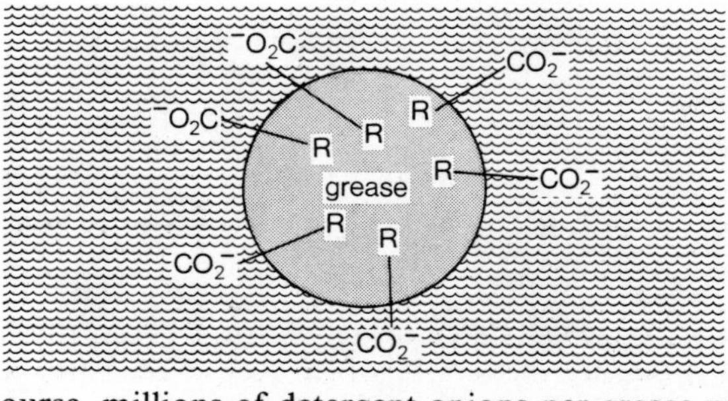

There are, of course, millions of detergent anions per grease particle.

* The significance of the prefix *cis-* is explained on page 137.

Catalytic hydrogenation of fats at high pressures and temperatures leads to reduction of them to long chain alcohols. Thus glyceryl trilaurate, which occurs in coconut oil, palm kernel oil, etc., gives dodecan-1-ol (lauryl alcohol) ($C_{12}H_{25}OH$). These long chain alcohols can be sulphated with sulphuric acid to give hydrogensulphate esters, themselves acids whose sodium salts are detergents.

$$C_{12}H_{25}OH + H_2SO_4 \longrightarrow C_{12}H_{25}O.SO_2.OH \xrightarrow{NaOH} C_{12}H_{25}O.SO_2O^-Na^+$$

The disadvantage of soaps is that the calcium salts of the carboxylic acids are insoluble, so that they give scums when used in hard water which contains calcium ions. The synthetic detergents do not suffer from this disadvantage because the calcium salts of the alkyl hydrogensulphates are soluble, so scums are not formed.

Polyesters

Esterification of a dihydric alcohol with a dibasic acid can give a long-chain '*polyester*'. The most useful of these, as a synthetic fibre, is 'Terylene' or 'Dacron', the polyester of ethane-1,2-diol (ethylene glycol) and the dibasic aromatic acid benzene-1,4-dicarboxylic acid (terephthalic acid).

$$HOCH_2CH_2OH + HO_2C-C_6H_4-CO_2H + HOCH_2CH_2OH + \cdots + HO_2C-C_6H_4-CO_2H + HOCH_2CH_2OH$$

$$\downarrow$$

$$HOCH_2CH_2O.CO-C_6H_4-CO.OCH_2CH_2O\cdots\cdots CO-C_6H_4-CO.OCH_2CH_2OH$$

Very long chains indeed are built up. The molecules are linear, and fibres with very desirable physical properties are formed.

Acid (acyl) halides

These are prepared by the action of halogenating agents on carboxylic acids (page 121). Acid chlorides are the most common.

Because of the carbonyl polarization, the carbonyl carbon is highly susceptible to attack by nucleophiles, and because the halogen is a good leaving group (as Hal^-), nucleophilic substitution at the carbonyl carbon is an easy process. Thus acyl halides are very readily hydrolyzed:

$$\underset{H_2\ddot{O}:}{R-\overset{\delta+}{C}(=\overset{\delta-}{O})-Cl} \xrightarrow{slow} R-\underset{+OH_2}{\overset{\bar{O}}{C}}-Cl \xrightarrow{fast} R-\overset{O}{\overset{\|}{C}}-\overset{+}{O}H_2 + Cl^- \xrightarrow[fast]{-H^+} RC(=O)OH$$

This reaction takes place rapidly at room temperature, and acid chlorides give fumes of hydrogen chloride in moist air. Acetyl chloride (ethanoyl chloride), for example, is a volatile, fuming, pungent-smelling liquid (b.p. 52 °C).

Acid halides react easily with other nucleophiles also. Alcohols give esters by a similar mechanism:

$$RCO.Cl + R'OH \longrightarrow RCO.OR' + HCl,$$

ammonia gives amides:

$$RCO.Cl + 2NH_3 \longrightarrow RCO.NH_2 + NH_4Cl,$$

and carboxylate anions (usually as the sodium salts) give anhydrides:

$$RCO.Cl + RCO.O^- \longrightarrow RCO.O.COR + Cl^-$$

The mechanisms of all these reactions are analogous to that given in detail for hydrolysis, and can easily be deduced using the appropriate nucleophiles in place of the water molecule.

Acid chlorides, in solution in boiling xylene, can be reduced to aldehydes with hydrogen in the presence of palladium on barium sulphate as catalyst. This is known as the 'Rosenmund' reaction (cf. Chapter 10, page 103).

$$RCO.Cl + H_2 \longrightarrow RCHO + HCl$$

Anhydrides

These can be prepared by dehydration of the acids (page 121) or by treatment of the sodium salt of the acid with the acid chloride, as above.

Anhydrides react with nucleophiles for the same reason as acid halides, although they are rather less reactive because carboxylate anions are not such good leaving groups as halide ions. Thus, for example, acetic anhydride (ethanoic anhydride) (b.p. 139.5 °C) does not fume in moist air, and is hydrolysed slowly in water but rapidly by alkali.

$$\underset{H_2O}{R{-}\overset{\overset{\large O}{\|}}{C}{-}O.CO.R} \longrightarrow R{-}\underset{\underset{^{+}OH_2}{|}}{\overset{\overset{\large O}{\|}}{C}} + RCO.O^- \longrightarrow 2RCO.OH$$

Anhydrides react with alcohols to give esters:

$$RCO.O.CO.R + R'OH \longrightarrow RCO.OR' + RCO.OH,$$

and with ammonia to give amides:

$$RCO.O.COR + NH_3 \longrightarrow RCO.NH_2 + RCO.OH$$

These reactions are mechanistically similar to the analogous reactions of acid halides but with the carboxylate, instead of the halide, anions as leaving groups.

Amides

Amides can be prepared by reactions of ammonia with esters, acid halides or anhydrides, or by heating the ammonium salts of carboxylic acids:

$$RCO.ONH_4 \xrightarrow{\text{heat}} RCO.NH_2 + H_2O$$

[The amide of carbonic acid [$CO(OH)_2$] is called *urea* [$CO(NH_2)_2$]].

Most amides are solids, and the lower members are water-soluble.

Once again, they can be hydrolysed, slowly by water, rapidly on heating with alkali, by mechanisms analogous to those of the hydrolysis of the other acid derivatives, and involving nucleophilic attack on the carbonyl carbon.

$$RCO.NH_2 + H_2O \longrightarrow RCO_2H + NH_3$$

Amides are only very feebly basic, although they can be regarded as substituted ammonias. It is the unshared pair of electrons which is the source of the basicity of ammonia:

$$H_3\ddot{N} + H^+ \longrightarrow H_4N^+$$

This unshared pair in amides is rendered less available for co-ordination as above by the electron attraction of the carbonyl group, and the resonance:

$$R{-}C({=}O){-}\ddot{N}H_2 \longleftrightarrow R{-}C(O^-){=}\overset{+}{N}H_2$$

This kind of resonance is present in all these acid derivatives:

$$R{-}C({=}O){-}\ddot{X} \longleftrightarrow R{-}C(O^-){=}X^+$$

and is at least partly responsible for the low reactivity of the carbonyl groups in esters, acid halides, and anhydrides as well as in amides.

Thus amides are hardly basic at all: indeed the electron attraction of the carbonyl group enables one of the hydrogens attached to the nitrogen to separate as a proton in the right environment. In other words, amides can act as very weak *acids*.

$$R{-}C({=}O){-}NH{-}H \rightleftharpoons R{-}C({=}O){-}\ddot{N}H^- + H^+$$

$$\updownarrow$$

$$R{-}C(O^-){=}NH$$

Thus amides can sometimes form metal salts.

They undergo two other important reactions. With nitrous acid [sodium nitrite (systematically nitrate (III)) and dilute hydrochloric acid] in the cold they give carboxylic acids and nitrogen, the $-NH_2$ group being replaced by $-OH$:

$$RCO.NH_2 + HNO_2 \longrightarrow RCO.OH + N_2 + H_2O$$

On heating with bromine and potassium hydroxide they undergo the *Hofmann* reaction, giving *primary amines* (Chapter 12, page 143).

$$RCO.NH_2 \xrightarrow[\text{heat}]{Br_2/KOH} \underset{\text{primary amine}}{RNH_2}$$

Although the mechanisms of both these reactions are reasonably well understood, discussion of them would be out of place here.

The amine obtained in the Hofmann reaction has one carbon fewer than the amide. The amine with the same number of carbon atoms as the amide is formed from the latter by reduction (e.g. with sodium and ethanol).

$$RCO.NH_2 \xrightarrow{Na/EtOH} RCH_2NH_2$$

Finally, amides can be dehydrated with phosphorus pentoxide to give *cyanides* (or *nitriles*) (Chapter 12, page 130).

$$RCO.NH_2 \xrightarrow{-H_2O} \underset{\text{nitrile}}{RCN}$$

Polyamides

'Polyamides' can be formed by loss of water between diamines and dicarboxylic acids. For example, 1,6-diaminohexane and hexanedioic acid (adipic acid) give the familiar polymer *Nylon-66*.

$$\underset{\text{1,6-diaminohexane}}{H_2N(CH_2)_6NH_2} + \underset{\text{hexanedioic acid (adipic acid)}}{HO_2C(CH_2)_4CO_2H} + \text{--------------}$$

$$+ H_2N(CH_2)_6NH_2 + HO_2C(CH_2)_4CO_2H$$

$$\downarrow$$

$$H_2N(CH_2)_6NH.CO(CH_2)_4CO. \text{--------------} NH(CH_2)_6NH.CO(CH_2)_4CO_2H$$

Nylon-66

Each amide linkage $-NH.CO-$ or $-CO.NH-$ is formed with loss of a molecule of water, and a very long linear polymer is thus formed.

Hydroxy-acids : optical isomerism

Certain hydroxy-acids are important natural products, and display a particular and important kind of isomerism, namely *optical isomerism*, which arises from the arrangement in space of the groups in the molecule. It is therefore one type of *stereoisomerism*.

The simplest of these hydroxy-acids, and the only one to be considered in any detail here, is 2-hydroxypropanoic acid, $CH_3CH(OH)CO_2H$ (usually known as lactic acid). Others are 2-hydroxybutanedioic acid (malic acid):

$$\begin{array}{l} CH(OH)CO_2H \\ | \\ CH_2CO_2H \end{array}$$

which occurs in unripe apples, and 2,3-dihydroxybutanedioic acid (tartaric acid):

$$\begin{array}{l} CH(OH)CO_2H \\ | \\ CH(OH)CO_2H \end{array}$$

which occurs as a salt in the precipitate (tartar) formed in wine vats.

Lactic acid was isolated in 1780 by Scheele from sour milk, hence its common name. It had m.p. 18 °C. In 1808 Berzelius isolated a lactic acid from mammalian muscle, in which it is formed by metabolic processes. Although apparently chemically identical with Scheele's acid, it had m.p. 26 °C, and displayed the phenomenon of *optical activity*, which was not shown by Scheele's acid. Berzelius therefore called it 'sarcolactic acid' to distinguish it from the acid obtained from milk.

Optical activity is the ability of a crystal or solution to rotate the *plane of polarization* of *plane polarized light*. Ordinary light consists of vibrations in all directions perpendicular to the direction in which the light is travelling. For example, for light travelling in a direction perpendicular to the plane of the paper, vibrations are in all directions in that plane:

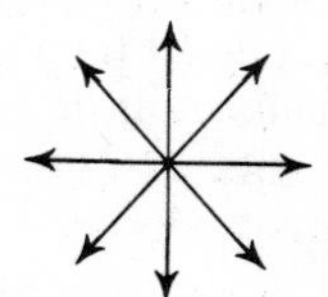

Plane polarized light consists of vibrations in *one direction only*, e.g.

and that direction is the *plane of polarization*. It is produced when light is passed through certain crystals, such as Iceland spar, which exhibit double refraction. The single light ray is split into two rays, known as *ordinary* and *extraordinary*, which are both polarized in planes perpendicular to one another, and which have different refractive indices.

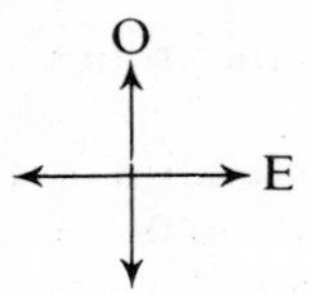

A single plane polarized beam can be obtained by passing ordinary light through a crystal of Iceland spar which has been cut through diagonally, so that it contains a crystal–air interface. The ray with the lower refractive index is then removed by total internal reflection at the interface. This arrangement is called a *Nicol prism* (Figure 11.1).

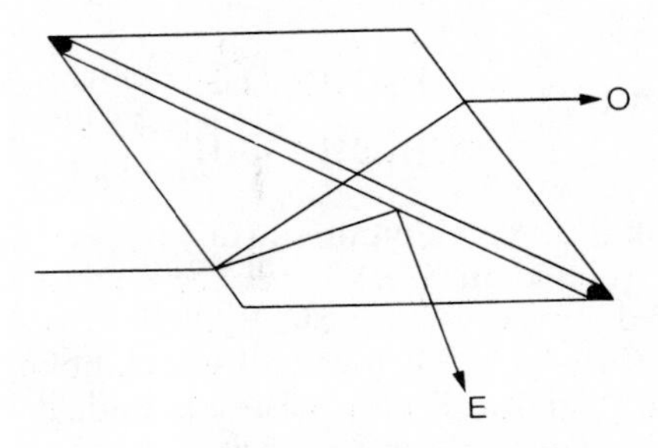

Figure 11.1

If a second Nicol prism is placed in line so that the ordinary ray passes through it, and turned about the axis of the instrument (the direction of the light) through 90°, the ray which was ordinary with respect to the first Nicol becomes extraordinary with respect to the second, and is extinguished by total internal reflection.

Now, if a tube containing the solution of an *optically active substance* (one which displays optical activity) is placed between the two Nicols, the plane of polarization is rotated, and the *angle of rotation* is the same as the angle through which the second Nicol must be rotated until the light is again extinguished. This arrangement, shown schematically in Figure 11.2, is the basis of the *polarimeter*, the instrument used to measure angles of rotation.

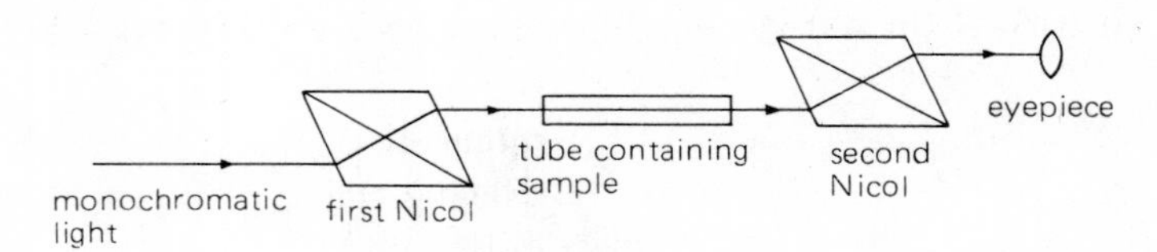

Figure 11.2

The optical activity of *crystals* which display it (e.g. quartz) arises from lack of symmetry in the arrangement of the molecules in the crystal. The optical activity of *solutions* arises from lack of symmetry in the molecules of the optically active solute itself. In order for such a substance to be optically active, the criterion is that its molecules must be *dissymmetric*, i.e. *they must display such a lack of symmetry as would render the molecule and its image in a mirror not capable of superimposition on one another*. The term *asymmetric* (lacking any symmetry at all) used to be used in this context but for reasons which are rather subtle, and in any case not relevant to this simple account, it is not quite accurate, and the term dissymmetric is now preferred. Dissymmetric objects and their (non-superposable) mirror images are called *enantiomers*, and they are quite common. For example, a right and a left hand are mirror images, and are not superposable (try it!) and are therefore enantiomers, as also are a left and a right-handed screw or helix.

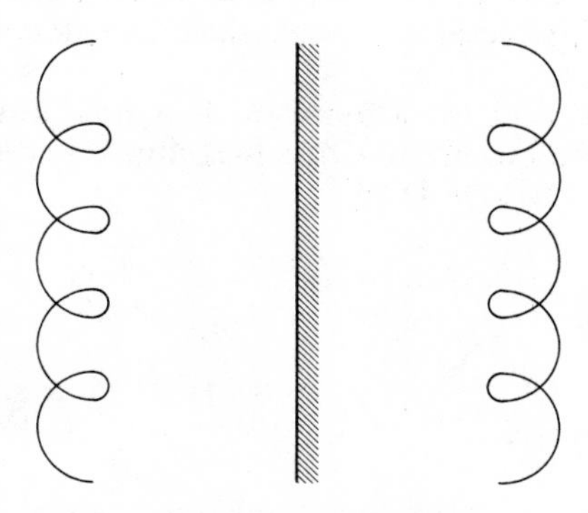

Figure 11.3

The right- or left-handedness of enantiomers is called their *chirality*.

Many molecules are enantiomeric, and this chirality can arise from various structural features. The most common of these features, and the only one which need concern us in this account, is the presence of a single *asymmetric carbon atom*, i.e. a carbon atom to which *four different atoms or groups are attached*. This was first realized by van't Hoff and le Bel in 1874, and was the experimental basis which led them both to postulate the tetrahedral distribution of the valencies of carbon, because such a distribution is necessary if a molecule with a carbon atom attached to four different groups is to be asymmetric. Such a molecule and its mirror image are not superposable, as can easily be seen with the aid of models (the student *must himself* try this with models).

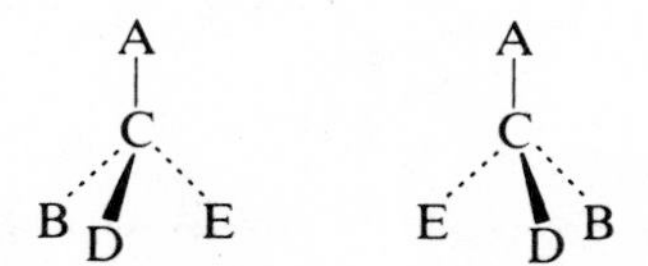

(In these formulae the atom A, whose bond to the central carbon is shown as a full line, lies in the plane of the paper; atom D, whose bond to the central carbon is wedge-shaped, is above that plane; and atoms B and E, whose bonds to the central carbon are shown as broken lines, lie below that plane.)

Since the object and its mirror image are not superposable they are *different*, but only in this rather subtle sense. In fact the only difference is that one will rotate the plane of polarization of plane polarized light to the right [the (+)-form], and the other equally to the left [the (−)-form]. They will have essentially the same physical and chemical properties apart from this.

A 50–50 mixture of the two enantiomers, which will not be optically active because the rotations of the two forms cancel each other out, need not, however, have the same melting point (for example), because of the physical effects of the components of a mixture on each other's melting points. Such a mixture is called a *racemic mixture* or *racemate* and it may have a melting point either higher or lower than that of the enantiomers, depending on the precise nature of the interactions between the molecules of the enantiomers in the crystalline racemate.

Returning to lactic acid for a moment, it is now obvious that Scheele's acid (from milk) is the racemate, while Berzelius's acid (from muscle) is an enantiomer [the (+)-form, in fact].

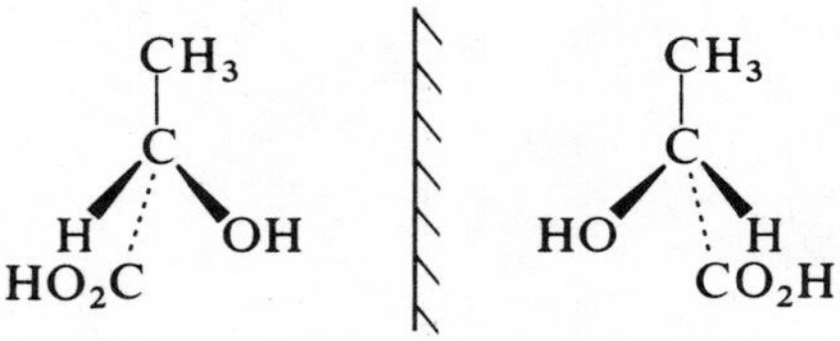

CH₃ on C; H, OH, HO_2C | mirror | CH₃ on C; HO, H, CO_2H

In order conveniently to write such three dimensional formulae in a plane, the so-called *Fischer projection* formulae are used. The convention is that groups lying in or behind the plane of the paper (normally the main chain of the molecule) are projected vertically and those lying in front of that plane are projected horizontally. Thus the above formulae of the enantiomeric lactic acids are written in Fischer projection below.

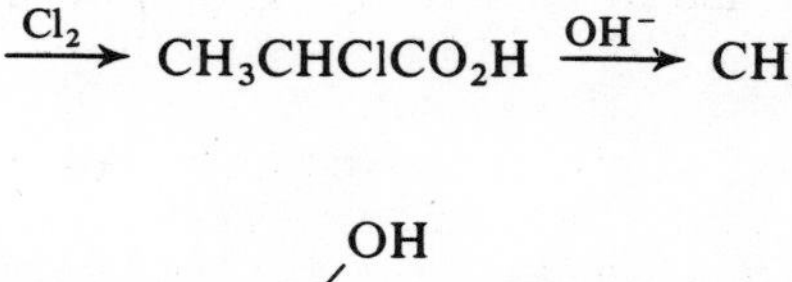

$$\begin{array}{c} CH_3 \\ | \\ H-C-OH \\ | \\ CO_2H \end{array} \qquad \begin{array}{c} CH_3 \\ | \\ HO-C-H \\ | \\ CO_2H \end{array}$$

Lactic acid can be synthesized by the following routes.

$$\underset{\text{propanoic acid}}{CH_3CH_2CO_2H} \xrightarrow{Cl_2} CH_3CHClCO_2H \xrightarrow{OH^-} \underset{\text{lactic acid}}{CH_3CH(OH)CO_2H}$$

$$\underset{\text{ethanal}}{CH_3CHO} \xrightarrow{HCN} CH_3CH\begin{matrix} OH \\ CN \end{matrix} \xrightarrow{OH^-} \underset{\text{lactic acid}}{CH_3CH(OH)CO_2H}$$

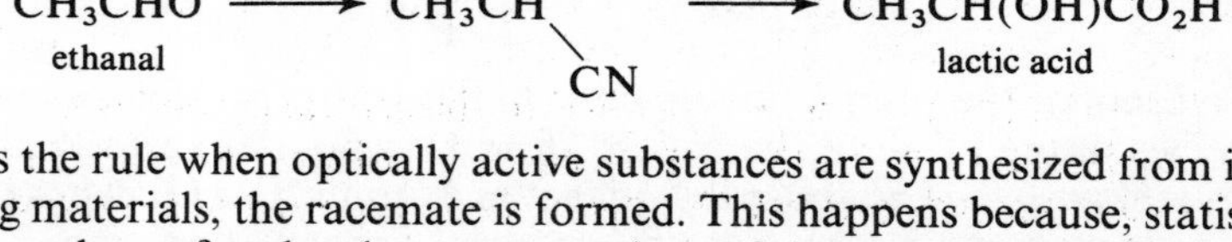

As is the rule when optically active substances are synthesized from inactive starting materials, the racemate is formed. This happens because, statistically, equal numbers of molecules, atoms, or ions of the new reagent whose reaction produces the asymmetric carbon atom approach the atom with which they react from both possible directions, so that equal amounts of the two enantiomers are formed. Thus, for example, in the first stage of the second of the above reactions the following events are equally probable.

$H_3C(H)C{=}O + CN^- \longrightarrow H_3C(H)C(CN)O^-$

$H_3C(H)C{=}O + CN^- \longrightarrow H_3C(H)C(O^-)CN$

enantiomers

The conversion of an enantiomer into a racemate (i.e. the conversion of half of it into the other enantiomer to give an equilibrium mixture) is called *racemization*. Lactic acid can be racemized by boiling it with alkali.

The reverse process, that of obtaining an enantiomer from the racemic mixture is called *resolution*, and is of great importance. *Pasteur* (1848–1858) devised the three classical methods of resolution.

1. The crystals of the enantiomers of the substance, or of one of its salts if it is an acid, are often not isomorphous (i.e. do not have the same crystalline form), being non-superposable mirror images of one another. They can therefore sometimes be separated mechanically, or when seed crystals of one enantiomer have been obtained, they can be allowed to grow in a supersaturated solution of the racemate. Only that enantiomer will grow on these crystals since the other is not isomorphous with it. This method is now only of historical interest.
2. A suitable bacterium or mould, when grown in a solution containing the racemate, may consume one enantiomer much more rapidly than the other. This other enantiomer is thus ultimately left behind. Pasteur used the mould *penicillium glaucum* to resolve ammonium tartrate in this way.
3. The best of these methods is to form the salt of the racemic acid with one enantiomer of an optically active base. (A number of such bases occur naturally in optically active forms, e.g. quinine, cinchonine, strychnine.) The two salts formed from a racemic acid [(±)A] and, say, a base (+)B are (+)A(+)B and (−)A(+)B. These are no longer enantiomers, but differ from one another more fundamentally, as do, for example, the *combinations* formed by a *right* hand grasping another person's *right* hand, and a *right* hand grasping another person's *left* hand. Such combinations are known as *diastereomers*, and diastereomeric salts have physical properties (in particular, solubilities) different from one another, whereas the enantiomeric (+)- and (−)- acids do not. Thus, the salts can be separated by fractional crystallization, and the free acids then liberated by treatment with strong acids.

Racemic mixtures of optically active bases can be separated by an analogous method involving salt-formation with one enantiomer of an optically active acid. Several such acids are available in active forms, e.g. tartaric acid, camphorsulphonic acid.

It is because their resolution by this method is relatively easy that carboxylic acids have been the subjects of a great deal of the study which has been made of optical isomerism. Thus, while compounds such as that below, for example, can exist in enantiomeric forms, their separation from the racemic mixture which must from any synthesis from optically inactive starting materials is more difficult.

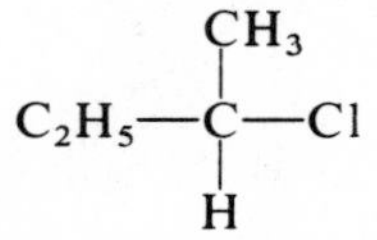

$$C_2H_5—\underset{\underset{H}{|}}{\overset{\overset{CH_3}{|}}{C}}—Cl$$

2-chlorobutane

However, such halo-compounds can be obtained in their optically active forms, and the stereochemical consequences of their nucleophilic substitution reactions are interesting and important, being dependent upon the mechanism of the substitution. S_N2 reactions lead to the *inversion* of the configuration of every molecule which reacts, because approach of the reagent from the 'back' of the molecule leads to a more stable transition state than approach from the front, and is therefore the course followed. Thus S_N2 reaction of 2-chlorobutane with potassium acetate (KOAc) in ethanol as solvent proceeds as follows:

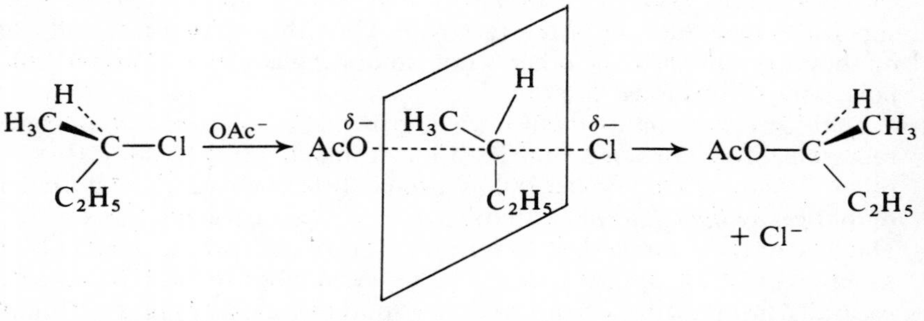

$$\text{H(H}_3\text{C)(C}_2\text{H}_5\text{)C—Cl} \xrightarrow{OAc^-} \overset{\delta-}{\text{AcO}}\cdots\text{C(H)(H}_3\text{C)(C}_2\text{H}_5\text{)}\cdots\overset{\delta-}{\text{Cl}} \longrightarrow \text{AcO—C(H)(CH}_3\text{)(C}_2\text{H}_5\text{)} + Cl^-$$

In the transition state, the three groups CH_3, C_2H_5, and H lie in a plane perpendicular to the plane of the paper and, during the course of the reaction, the molecule turns 'inside out' like an umbrella in a high wind. This inversion is called 'Walden inversion' after its discoverer.

The typical steroechemical result of S_N1 reactions of such optically active substances is *racemization*. This occurs because the carbonium ions which are intermediates in S_N1 processes are planar, since the central carbon atom is sp^2 hybridized (cf. Chapter 6, page 46). In the second stage of the reaction the nucleophile may attack from the front or the back of the carbonium ion, leading to the formation of both enantiomers, e.g.

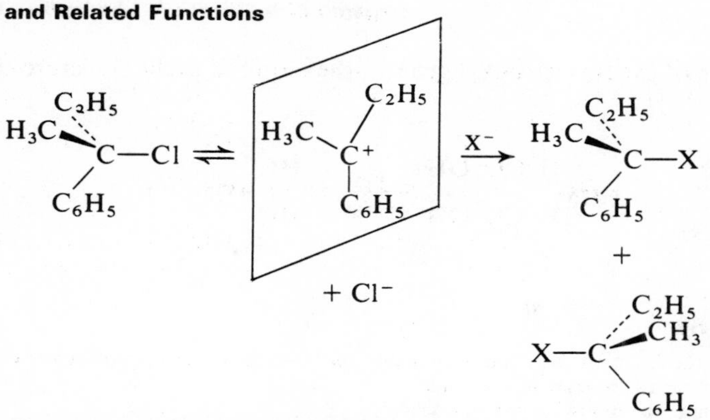

$$H_3C(C_2H_5)(C_6H_5)C{-}Cl \rightleftharpoons [H_3C(C_2H_5)(C_6H_5)C^+] + Cl^- \xrightarrow{X^-} H_3C(C_2H_5)(C_6H_5)C{-}X + X{-}C(C_2H_5)(CH_3)(C_6H_5)$$

In the absence of any complications, the two enantiomers are formed in equal yields, leading to complete racemization. Such simple behaviour is, however, rare; more usually, there is a complication of one kind or another, so that racemization is partial, being accompanied by some inversion or some retention of configuration.

These stereochemical consequences, namely complete inversion of configuration for S_N2 and (usually partial) racemization for S_N1, can be used to diagnose the mechanisms of reactions such as solvolyses for which kinetics do not give an unequivocal answer (cf. Chapter 8, page 80).

Maleic and fumaric acids : geometrical isomerism

The unsaturated dicarboxylic acids *cis*- and *trans*-butenedioic acids (maleic and fumaric acids) afford a classic example of geometrical isomerism (Chapter 8, page 80).

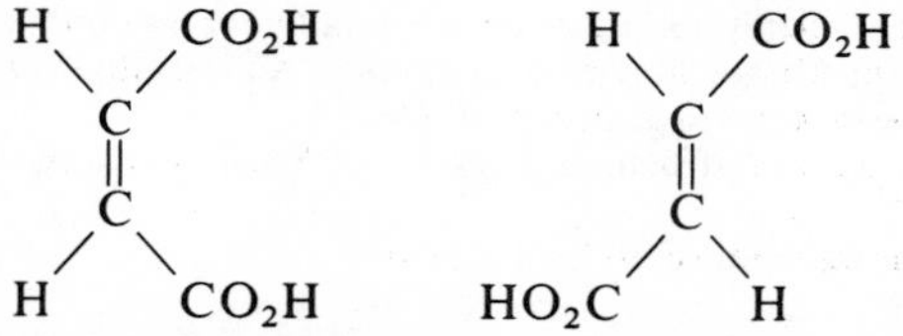

$$(H)(CO_2H)C{=}C(H)(CO_2H) \qquad (H)(CO_2H)C{=}C(HO_2C)(H)$$

cis-butenedioic acid (maleic acid) m.p. 130 °C

trans-butenedioic acid (fumaric acid) m.p. 287 °C

The meaning of the prefixes *cis*- and *trans*- is obvious from the above formulae.

Geometrical isomers are, of course, not enantiomers. There is no optical activity since such planar molecules cannot be dissymmetric. The plane of the molecule is itself a plane of symmetry and mirror images of either isomer are superposable on the object. The two geometrical isomers are *not* mirror images of one another. Thus both physical and chemical properties are different. For example, the pK_{a_1} values (for the first dissociation) are 1.92 for the *cis*- and 3.02 for the *trans*- acid, and the melting points and solubilities are widely different.

A striking chemical difference is that the *cis*- acid readily forms an anhydride internally, whereas the *trans*- acid does not, because, owing to the wide separation of the two carboxyl groups, the required cyclic structure cannot be formed.

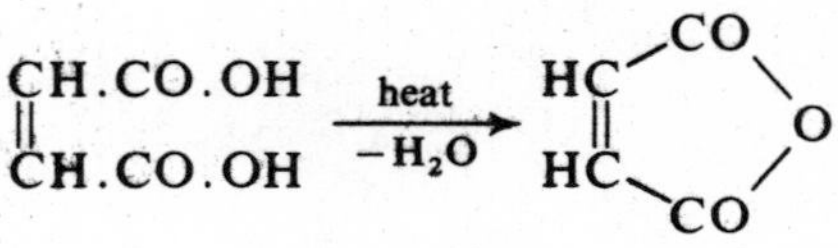

$$\begin{array}{l} CH.CO.OH \\ \| \\ CH.CO.OH \end{array} \xrightarrow[-H_2O]{heat} \begin{array}{l} HC-CO \\ \| \qquad\quad O \\ HC-CO \end{array}$$

Questions

1. (*a*) Give the equation and conditions of any *one* reaction for the preparation of ethanol from readily available starting materials.
   (*b*) Examine the following reaction scheme.

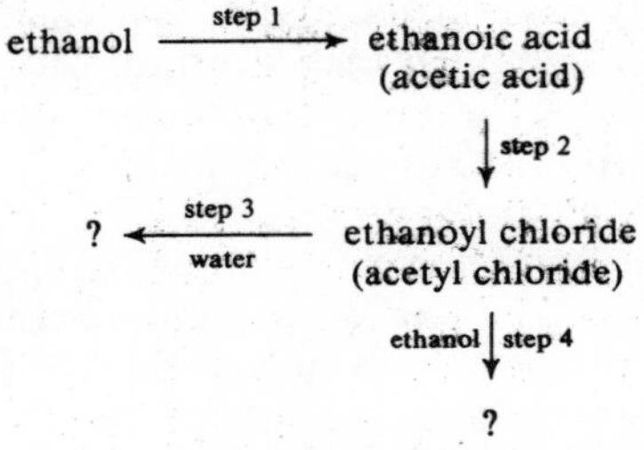

ethanol —step 1→ ethanoic acid (acetic acid)
ethanoic acid (acetic acid) —step 2→ ethanoyl chloride (acetyl chloride)
ethanoyl chloride (acetyl chloride) —step 3, water→ ?
ethanoyl chloride (acetyl chloride) —step 4, ethanol→ ?

   (i) Name a suitable reagent or reagents for step 1 and give the conditions required.
   (ii) Name a suitable reagent for step 2 and give the conditions required.
   (iii) Name the products of step 3.
   (iv) Give a structural formula for the organic product of step 3.
   (v) Name the products of step 4.
   (vi) Give a structural formula for the organic product of step 4.

   (*c*) 10 Millimoles of ethanoyl chloride were added gradually to 20 millimoles of phenylamine (aniline). On subsequently adding water, a crystalline compound **X** was precipitated. It was filtered off, and the filtrate allowed to evaporate. A crystalline residue **Y** remained, which was freely soluble in water to give an acid solution.
   (i) Give the names and structural formulae of **X** and **Y**, and write an equation for their formation.
   (ii) Why was the aqueous solution of **Y** acid?

(NISEC, 1973)

2. (*a*) Write out and label appropriately the *cis* and *trans* forms of the compound, ClCH:CHCl which we shall call *X*.
   (*b*) Explain why *X* is able to form geometric isomers.
   (*c*) Write the structural formula of a compound isomeric with *X* which cannot form geometric isomers.
   (*d*) Write the structural formula of the product, *Y*, formed when excess bromine is added to *X*.
   (*e*) Each asymmetric carbon grouping in *Y* can be considered to contribute an optical rotation of $+a°$ or $-a°$ to the rotation caused by the whole molecule.
   (i) Explain how *Y* can exist as two optically active molecules and one optically inactive molecule.
   (ii) What will be the resultant optical rotation of the two active forms?
   (iii) Explain how another optically inactive form of *Y* can be obtained from the two optically active forms.

   (*f*) (i) Write the electronic structure of the carbon atom using *s*, *p*, *d*, etc. notation.
   (ii) Explain how a tetrahedral distribution of C–H bonds arises in methane using this electronic structure and the idea of orbitals.

(SUJB, 1975)

3. What do you understand by *geometrical isomerism* and by *optical isomerism*?
   Give suitable examples in support of your answer and show clearly the molecular structures associated with **each**. What optical property is referred to in *optical isomerism*?
   What is meant by a *racemic mixture*? Describe and explain **one** chemical method for the resolution of a racemic mixture of an acid displaying optical isomerism.

(WJEC, 1975)

4. The structures of ethers, esters and anhydrides are similar in that each consists of two alkyl and/or acyl groups attached to an oxygen atom. Compare and contrast in tabular form the methods of preparation, physical properties and chemical reactions of an ether, an ester and an acid anhydride.

(London, 1973)

5. (*a*) Name the reagents required to prepare ethanoyl (acetyl) chloride, state the conditions required and give the equation for the reaction.
   (*b*) In each case, state the type of reaction and outline the mechanism by which ethanoyl chloride reacts with (i) water, (ii) benzene.
   (*c*) Calculate the volume of 2M (2N) NaOH which is required to neutralize the products of the action of water on 0.05 mol ethanoyl chloride, naming the indicator you would use in the titration.

(JMB, 1974)

Chapter 12

Nitrogen-containing Functions

The cyano-function

Alkyl cyanides, R–C≡N, are sometimes also called *nitriles* of either the acid or the alkane with the same number of carbon atoms. Thus for example

$$CH_3C{\equiv}N$$

methyl cyanide, acetonitrile, or ethanonitrile
(b.p. 81 °C)

Aromatic nitriles such as $C_6H_5C{\equiv}N$ (benzonitrile or benzenecarbonitrile, b.p. 191 °C) also exist.

Isonitriles ($R{-}\overset{+}{N}{\equiv}\overset{-}{C}$) are isomeric with the nitriles, but will not be discussed further here.

Formation of the cyano-function

Nitriles with the cyano-group attached to sp^3 carbon are usually prepared by heating a haloalkane (usually the iodo-compound) with sodium or potassium cyanide in ethanolic solution. The cyanide ion is a good nucleophile, and replacement of halogen occurs by the S_N1 or S_N2 mechanism depending on the halide and the conditions (cf. Chapter 8, page 81).

$$RI + CN^- \longrightarrow RCN + I^-$$

Dehydration of amides also give nitriles (Chapter 11, page 130).

$$RCO.NH_2 \xrightarrow[(P_2O_5)]{-H_2O} RCN$$

Aromatic nitriles can be prepared by this latter method and also by treatment of diazonium salts with sodium or potassium cyanide (page 149).

Reactions of the cyano-function

The most important reactions of nitriles are hydrolysis and reduction.

Acid hydrolysis gives the ammonium salt of a carboxylic acid, *via* the amide.

$$RCN \xrightarrow{H_2O} RCO.NH_2 \xrightarrow{H_2O} R.CO.O^-NH_4^+$$

Alkaline hydrolysis gives, as would be expected, the alkali metal salt of the same carboxylic acid and ammonia.

$$RCN \xrightarrow{NaOH} RCO.O^-Na^+ + NH_3$$

Hydrolysis of nitriles is often a useful way of introducing a carboxyl group (cf. Chapter 12, page 118).

Nitriles are readily reduced with, for example, sodium and ethanol, to *primary amines* (cf. page 143).

$$RCN \xrightarrow[(Na/EtOH)]{4H} RCH_2NH_2$$

a primary amine

These reactions are undergone by both aliphatic and aromatic nitriles, and afford methods of introducing an extra carbon atom into a chain, e.g.

$$ROH \xrightarrow{P/I_2} RI \xrightarrow{CN^-} RCN \xrightarrow{\text{hydrolysis}} RCO_2H \xrightarrow{LiAlH_4} RCH_2OH$$

The nitro-function

Nitro-compounds contain the nitro-group, $-NO_2$

$$\left(-\overset{+}{N}(\overset{-}{O})=O \longleftrightarrow -\overset{+}{N}(=O)\overset{-}{O} \right)$$

This resonance is similar to that of the carboxylate ion group (Chapter 11, page 119). Since the nitro-group is dipolar, and is linked to the rest of the molecule by its positive end, it is strongly electron-attracting, i.e. it has a powerful $-I$ effect.

Nitroalkanes are isomeric with alkyl nitrites (R–O–N=O), which are simply esters of nitric(III) acid (nitrous acid) and have the expected properties of esters. The simplest members are:

CH_3NO_2
nitromethane (b.p. 101 °C)

$C_6H_5NO_2$
nitrobenzene (b.p. 211 °C)

Nitro-compounds can be obtained by direct nitration. Nitration of alkanes (Chapter 5, page 44) is a homolytic reaction, while that of arenes (Chapter 7, pages 61–8) is the classic example of electrophilic substitution in arenes. Other methods are also available for the preparation of nitroalkanes, such as the treatment of haloalkanes with silver nitrite (cf. Chapter 8, page 79).

Reactions of the nitro-function

The most important reaction of the nitro-function is its reduction to the primary amino-group. Acidic reducing agents (e.g. tin and concentrated hydrochloric acid) are used.

$$-NO_2 \xrightarrow[(Sn/HCl)]{6H} -NH_2 + 2H_2O$$

This is by far the most useful way of preparing aromatic amines (cf. page 143), e.g.

NO_2 (nitrobenzene) $\xrightarrow{Sn/HCl}$ NH_2 (aniline)

nitrobenzene   aniline

In this case the aniline is volatile in steam, and is usually steam-distilled out of the reaction mixture after it has been made alkaline.

Because of the electron-attraction of the nitro-group, the aromatic nucleus is *deactivated* by a nitro-group already present towards further attack by electrophiles. Thus for example, further nitration of nitrobenzene requires more vigorous conditions. The effects of substituent groups on the positions at which further substitution takes place were discussed in Chapter 7 (pages 66–8). It was there seen that electron attracting groups like *nitro*- deactivate the *ortho*- and *para*- (2- and 4-) positions more than the *meta*- (3-position) so that electrophiles are 'directed' to the *meta*- (3-) position.

$\bar{O}$ + O N

$\delta+$

$\delta+$

The product of further nitration of nitrobenzene is therefore almost entirely *m*-(1,3-) dinitrobenzene.

NO_2 $\xrightarrow{\text{fuming } HNO_3/H_2SO_4 \text{ (100 °C)}}$ NO_2 NO_2

Amino-functions

Compounds containing the amino-function are called amines, and may be regarded as substituted ammonias. If one of the hydrogens of ammonia has been replaced by an organic group, the compound is a *primary amine*; if two, a *secondary amine*; and if three, a *tertiary amine*. Substituted ammonium salts exist in which all four hydrogens of the ammonium ion have been replaced by organic groups and are known as *quaternary ammonium salts*. (Note that the terms 'primary', 'secondary', and 'tertiary' have different meanings when applied to amines and to alcohols. Their meanings in the latter case were discussed in Chapter 9, page 86).

$R—NH_2$ primary amine

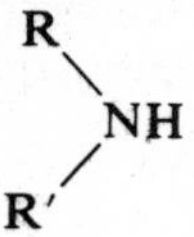

$R(R')NH$ secondary amine

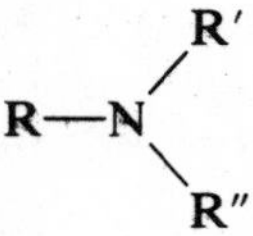

$R—N(R')(R'')$ tertiary amine

$R—N^+(R')(R'')—R'''$ X^- quaternary ammonium salt

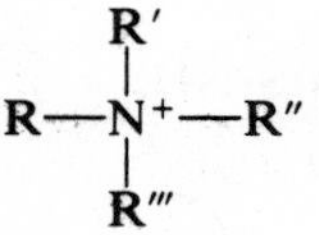

The organic groups need not, of course, all be different, and they may be aliphatic or aromatic. The lower aliphatic amines are gases (although higher members are liquids) with strong, 'fishy', ammoniacal odours. Aromatic amines are high-boiling liquids or solids, e.g.

CH_3NH_2
methylamine
(gaseous)

$(CH_3)_2NH$
dimethylamine
(b.p. 7 °C)

$(CH_3)_3N$
trimethylamine
(b.p. 3 °C)

$C_6H_5NH_2$
phenylamine, aminobenzene
(aniline)
(b.p. 183 °C)

(The systematic names phenylamine and aminobenzene are hardly ever used in practice, and this compound is almost always called aniline.)

Formation of amino-functions

Aliphatic and aromatic primary amines can be prepared by *reduction* of nitro-compounds (page 142), cyanides (page 141), or amides (Chapter 11, page 130), and by the *Hofmann reaction* from amides (Chapter 11, page 130).

Alkylation of ammonia with haloalkanes, in which ammonia acts as a nucleophile in replacing the halogen (Chapter 8, page 79), gives primary amines, but further alkylation of the primary amine also occurs and the reaction, being difficult to control, gives also secondary, tertiary, and quaternary products (cf. page 145).

$$RX \xrightarrow{NH_3} RNH_2 \xrightarrow{RX} \text{other products}$$

These product mixtures can be difficult to separate, although, while other more complex methods are sometimes available, it is often necessary to use this method to obtain the secondary and tertiary amines, and quaternary salts.

The chemistry of amino-functions

The properties and reactions of amines are dominated by the nitrogen atom's unshared pair of electrons which, as in ammonia, are available for bond formation by co-ordination with electron acceptors. For this reason amines, like ammonia, are *bases* and *nucleophiles*; e.g.

$$R{-}\ddot{N}H_2 \quad H^+ \longrightarrow R{-}\overset{+}{N}H_3$$

$$R_2\ddot{N}H \quad H^+ \longrightarrow R_2\overset{+}{N}H_2$$

$$R_3\ddot{N} \quad H^+ \longrightarrow R_3\overset{+}{N}H$$

Their basicity thus leads to the formation of salts with acids, in which amines are soluble. These salts can be obtained as crystalline solids. The amines can be regenerated from the salts by treatment with alkali in the cold, e.g.

$$\underset{\text{methylamine}}{CH_3NH_2} \underset{OH^-}{\overset{HCl}{\rightleftharpoons}} \underset{\substack{\text{methylammonium chloride} \\ \text{(methylamine hydrochloride)}}}{CH_3NH_3^+Cl^-}$$

$$\underset{\text{aniline}}{C_6H_5NH_2} \underset{OH^-}{\overset{HCl}{\rightleftharpoons}} \underset{\substack{\text{phenylammonium chloride} \\ \text{(aniline hydrochloride,} \\ \text{anilinium chloride)}}}{C_6H_5NH_3^+Cl^-}$$

The basic strength of amines depends on the *availability* of the unshared electrons on the nitrogen. Thus repulsion of electrons towards the nitrogen leads to an increase in basic strength. Conversely, attraction of electrons away from the nitrogen, or, more particularly, delocalization of the unshared pair in an aromatic nucleus, as in aromatic amines, is a base-weakening factor. Thus base strength increases from ammonia to primary to secondary amines because of the $+I$ effects of the alkyl groups, e.g.

$$NH_3 < CH_3NH_2 < (CH_3)_2NH$$

	NH_3	CH_3NH_2	$(CH_3)_2NH$
pK_b	9.27	10.62	10.77

(N.B. pK_b *increases* with increasing base-strength.) There is a somewhat unexpected decrease in base strength when a third alkyl group is introduced [$(CH_3)_3N$ has pK_b 9.80]. This may arise from a *steric* effect, i.e. the over-crowding of the central nitrogen atom by three alkyl groups may make it a little less easy for the fourth group to approach near enough to form its bond with the nitrogen.

Aromatic amines are much weaker bases than aliphatic amines (aniline, $C_6H_5NH_2$, has pK_b 4.58), and this undoubtedly arises from delocalization of the unshared electron pair by the resonance:

$$:NH_2\text{-}C_6H_5 \longleftrightarrow {}^{+}NH_2\text{=}C_6H_5^{-}\ (\text{ortho}) \longleftrightarrow {}^{+}NH_2\text{=}C_6H_5^{-}\ (\text{para})$$

Amines, like ammonia, are nucleophiles, although for the above reason aromatic amines are weaker nucleophiles than aliphatic amines. Thus amines replace the halogen in haloalkanes, resulting in *alkylation* of the amines. In this way a primary amine gives a secondary ammonium salt,

$$R{-}\ddot{N}H_2 + R'{-}X \longrightarrow R{-}\overset{+}{N}H_2{-}R' \quad X^-$$

although progressive alkylation often leads to the formation of primary, secondary, and tertiary amines as their salts, and quaternary ammonium salts; for example:

$$CH_3NH_2 \xrightarrow{CH_3I} \underset{\text{(as salt)}}{(CH_3)_2NH} \xrightarrow{CH_3I} \underset{\text{(as salt)}}{(CH_3)_3N} \longrightarrow (CH_3)_4N^+I^-$$

$$C_6H_5NH_2 \xrightarrow{CH_3I} C_6H_5NHCH_3 \xrightarrow{CH_3I} \text{etc.}$$

N-methylaniline
(*N*-methylphenylamine)
(as salt)

These, like other nucleophilic replacements, can take place by S_N1 or S_N2 mechanisms (cf. Chapter 8, pages 79–82).

Like ammonia, these nucleophilic amines can add to the carbonyl carbon of aldehydes and ketones. Loss of water from the addition products with primary amines (addition products cannot themselves normally be isolated, for this reason) then gives 'Schiff's bases' (cf. Chapter 10, page 110).

$$\overset{\delta+}{>C}{=}\overset{\delta-}{O} + R\ddot{N}H_2 \longrightarrow \left[>C(-O^-)(-\overset{+}{N}H_2R)\right] \longrightarrow \left[>C(-OH)(-NHR)\right] \xrightarrow{-H_2O} >C{=}NR$$

a Schiff's base

Because of the lack of hydrogen atoms on the nitrogen, Schiff's bases cannot, of course, be formed with secondary or tertiary amines.

A process similar to alkylation (*acylation*) leads to the formation of *substituted amides* by the use of acid chlorides and anhydrides, just as the corresponding reactions with ammonia gives simple amides (cf. Chapter 11, page 128).

Thus, with acid chlorides:

$$\underset{:NH_3}{-\overset{\delta+}{C}(Cl)=O^{\delta-}} \longrightarrow -C(Cl)(\overset{+}{N}H_3)-\ddot{O}^- \longrightarrow -C(=O)-\overset{+}{N}H_3 + Cl^- \xrightarrow{-H^+} \underset{\text{simple amide}}{-C(=O)-NH_2}$$

and:

$$\underset{R\ddot{N}H_2}{-\overset{\delta+}{C}(Cl)=O^{\delta-}} \longrightarrow -C(Cl)(\overset{+}{N}H_2R)-\ddot{O}^- \longrightarrow -C(=O)-\overset{+}{N}H_2R + Cl^- \xrightarrow{-H^+} \underset{\text{substituted amide}}{-C(=O)-NHR}$$

Similarly, with anhydrides:

$$\underset{:NH_3}{-\overset{\delta+}{C}(O.CO.R)=O^{\delta-}} \longrightarrow -C(O.CO.R)(\overset{+}{N}H_3)-\ddot{O}^- \longrightarrow -C(=O)-\overset{+}{N}H_3 + RCO.O^- \xrightarrow{-H^+} -C(=O)-NH_2$$

and:

$$\underset{R\ddot{N}H_2}{-\overset{\delta+}{C}(O.CO.R)=O^{\delta-}} \longrightarrow -C(O.CO.R)(\overset{+}{N}H_2R)-\ddot{O}^- \longrightarrow -C(=O)-\overset{+}{N}H_2R + RCO.O^- \xrightarrow{-H^+} -C(=O)-NHR$$

Substituted amides (acyl derivatives of amines) prepared thus are often crystalline solids with sharp melting points, and are used to characterize amines. Acetyl derivatives are often used for amines of higher relative molecular mass, and benzoyl derivatives for simpler amines; e.g.

$$C_6H_5NH_2 + (CH_3CO)_2O \longrightarrow \underset{\substack{N\text{-phenylacetamide} \\ \text{(acetanilide), m.p. 113 °C}}}{C_6H_5NH.CO.CH_3}$$

Substituted amides can also be formed by heating salts of amines with carboxylic acids, just as simple amides are formed by heating ammonium salts of carboxylic acids (cf. Chapter 11, page 000).

$$RCO.\bar{O}H_3\overset{+}{N}R' \xrightarrow[-H_2O]{\text{heat}} RCO.NHR'$$
substituted amide

Polyamides like Nylon (Chapter 11, page 130) are made in this way.

Amines undergo important and varied *reactions with nitric(III) acid* (*nitrous acid*) [in the form of sodium nitrate(III) (nitrite) and dilute acid, usually hydrochloric or sulphuric].

Aromatic primary amines, e.g. aniline, with sodium nitrite and dilute acid at 0–5 °C give solutions of *arenediazonium salts*; e.g.

$$C_6H_5NH_2 + HNO_2 + HCl \longrightarrow C_6H_5N_2^+Cl^- + 2H_2O$$
benzenediazonium chloride

The salts can be obtained as solids, but these are unstable and explosive, and the solutions of them, prepared as above, are usually used directly in further reactions (pages 149–50). One of their characteristic reactions is the so-called '*coupling*' reaction with phenols in alkaline solution, e.g.

$$C_6H_5{-}N_2^+ + C_6H_5OH \xrightarrow{NaOH} C_6H_5{-}N{=}N{-}C_6H_4{-}OH$$

The highly coloured products of such reactions are often dyes, many of which are used commercially. The phenol naphthalen-2-ol (usually known as β-naphthol) readily gives bright red dyes as coupling products with diazonium salts, and this reaction is used as a test for aromatic primary amines.

Aliphatic primary amines react with nitrous acid in the cold with copious evolution of a gas which usually consists largely of nitrogen. The alcohol in which the $-NH_2$ group has been replaced by $-OH$ is often formed, among other products, but the reaction is rather more complex than this. The difference in behaviour from that of the aromatic amines arises because, although diazonium ions are formed transiently, they are much less stable than the aromatic analogues; specifically they undergo rapid S_N1 reactions. Thus the alcohol can be formed if water is the nucleophile.

$$RNH_2 \longrightarrow RN_2^+ \longrightarrow R^+ + N_2$$
$$R^+ + H_2O \longrightarrow ROH + H^+$$

These processes are extremely rapid, because N_2, a very stable substance, is an excellent leaving group (cf. page 149), so that the diazonium ions have no more than a transient existence as reaction intermediates.

Products other than the simple alcohol are formed in many cases. For example, with methylamine the main product is methyl nitrite, CH_3ONO; and with 1-aminopropane ($CH_3CH_2CH_2NH_2$) propan-1-ol is formed in only about 7 per cent yield. Propan-2-ol and propene and, in the presence of halogen acids, 1- and 2-halopropanes are also formed. These products arise from other reactions of the carbonium ions. However, the copious evolution of a gas with cold nitrous acid is a characteristic test for aliphatic primary amines.

This reaction, together with the Hofmann reaction (Chapter 11, page 130), can be used in a sequence for the descent of an homologous series, whereby an alcohol is converted into the alcohol with one fewer carbon atom.

$$RCH_2OH \xrightarrow{\text{oxidation}} RCO_2H \xrightarrow{NH_3} RCO_2NH_4 \xrightarrow{\text{heat}}$$

$$RCO.NH_2 \xrightarrow{Br_2/NaOH} RNH_2 \xrightarrow{NaNO_2/HCl} ROH$$

Secondary amines do not give this reaction with nitrous acid, but instead undergo *nitrosation* on the nitrogen to give yellow, oily *nitrosamines*. This nitrosation reaction involves the *nitrosonium ion*, NO^+, as an intermediate. This is formed from nitric(III) acid (nitrous acid) in the presence of strong acids by processes analogous to those whereby the *nitronium ion*, NO_2^+, is formed from nitric(V) acid (nitric acid) in nitration (cf. Chapter 7, pages 61–3).

Nitrosation of secondary aliphatic and aromatic amines thus proceeds as follows.

$$HONO + H^+ \rightleftharpoons H_2\overset{+}{O}NO \rightleftharpoons H_2O + NO^+$$

$$R_2NH + NO^+ \rightleftharpoons R_2\overset{+}{N}(H)(NO) \xrightarrow{-H^+} R_2N{-}N{=}O$$

a nitrosamine

Tertiary aliphatic amines do not react with nitrous acid (apart from dissolving in the acidic medium to give the cation $R_3\overset{+}{N}H$) while *tertiary aromatic amines* undergo nitrosation in the nucleus to give solid nitroso-compounds, e.g.

$$C_6H_5N(CH_3)_2 \xrightarrow[(NO^+)]{HNO_2/H^+} 4\text{-}ON{-}C_6H_4N(CH_3)_2$$

N,N-dimethylphenylamine (dimethylaniline) → 4-nitroso-*N,N*-dimethylphenylamine (*p*-nitrosodimethylaniline) (a green solid)

This occurs readily, because the resonance mentioned earlier (page 145) by which the unshared electrons on the nitrogen in an aromatic amine are delocalized into the ring, means that amino-groups attached to an aromatic framework act as electron repelling groups. Thus the nucleus [like those of phenols (Chapter 9, page 96), and for a similar reason] is rendered electron-rich, and is therefore activated towards attack by electrophiles. Moreover, since it is the 2- and 4- (*ortho*- and *para*-) positions which receive the benefit of this extra electron density, it is at these positions that reaction most readily ensues.

Another example of this is the reaction of aniline with bromine water, in the cold, when an immediate precipitate of 2,4,6-tribromoaniline is formed.

$$C_6H_5NH_2 + 3Br_2 \longrightarrow \text{2,4,6-}Br_3C_6H_2NH_2 + 3HBr$$

It will be remembered (cf. Chapter 9, page 97) that phenol undergoes a similar rapid reaction, for a similar reason.

Diazonium salts

Solutions of diazonium salts are prepared (as mentioned on page 147), by treating a solution of a primary amine in an excess of dilute hydrochloric or sulphuric acid with a solution of sodium nitrite at 0–5 °C. Apart from the coupling reactions (page 147), there are many reactions in which the $-N_2^+$ group can be replaced by other atoms or groups.

These reactions are unusual in aromatic chemistry because some of them are nucleophilic, and proceed by S_N1 mechanisms.

The diazonium group is a hybrid of the following resonance forms, although other forms also make smaller contributions to the hybrid.

$$Ar{-}\overset{+}{N}{\equiv}N\!: \longleftrightarrow Ar{-}\ddot{N}{=}\overset{+}{N}\!:$$

Thus nitrogen and an aryl carbonium ion can be formed in the first stage of an S_N1 process.

$$Ar{-}N_2^+ \xrightarrow{\text{slow}} Ar^+ + N_2$$

It is the great stability of molecular nitrogen, making it an excellent leaving group (cf. page 147), which makes this process energetically worthwhile, even though aryl carbonium ions are rather unstable entities, and are not readily formed by processes analogous to the first stages of S_N1 reactions of, for example, tertiary haloalkanes. It needs this great stability of nitrogen to tip the energetic scales in favour of the formation of the aryl carbonium ion.

Once it is formed, Ar^+ can react with any available nucleophile, e.g. water, to give the phenol.

$$Ar^+ + H_2O \xrightarrow{\text{fast}} ArOH + H^+$$

Thus diazonium salts can be hydrolysed to phenols by boiling their (acidic) solutions.

$$ArNH_2 \xrightarrow{NaNO_2/\text{acid}} ArN_2^+ \xrightarrow{\text{boil}} ArOH$$

Halide ions can also act as nucleophiles, e.g. treatment of the diazonium salt solution with potassium iodide gives the iodo-arene. This is the best method of preparing these iodo-compounds (cf. Chapter 7, page 70).

$$ArN_2^+ + I^- \longrightarrow ArI + N_2$$

A cyano-group can also be introduced into an aromatic nucleus in this way (cf. page 140).

$$ArN_2^+ + CN^- \longrightarrow ArCN + N_2$$

However, to prepare chloro- and bromo- arenes, it is better to treat the diazonium salt (in solution) with copper(I) chloride (cuprous chloride) or bromide.

$$ArN_2^+ \xrightarrow{CuCl} ArCl + N_2$$

This is the '*Sandmeyer reaction*'. Its mechanism is not direct nucleophilic replacement of the $-N_2^+$ group as described above, but is a somewhat complex process involving aryl free radicals ($Ar\cdot$) as intermediates.

As well as those reactions mentioned above, there are many other reactions of the diazonium salts, which make them extremely valuable intermediates in aromatic chemistry. The following scheme summarizes the ways in which they can be used for introducing functional groups into the aromatic nucleus.

$$ArH \xrightarrow{\text{nitration}} ArNO_2 \xrightarrow{\text{reduction}} ArNH_2$$

$$ArNH_2 \xrightarrow{NaNO_2/\text{acid}} ArN_2^+X^-$$

$$ArN_2^+X^- \xrightarrow{CuBr} ArBr$$

$$ArN_2^+X^- \xrightarrow[\text{soln.}]{\text{boil acid}} ArOH$$

$$ArN_2^+X^- \xrightarrow{CuCl} ArCl$$

$$ArN_2^+X^- \xrightarrow{KCN} ArCN$$

$$ArN_2^+X^- \xrightarrow{KI} ArI$$

Questions

1. Describe with full practical details, a laboratory preparation of a *named* nitrile.
   Give the conditions for (a) the hydrolysis (b) the reduction, of the nitrile you have chosen, and name and give the structures of the products.
   What significance has the introduction of the –CN group into organic compounds?
   (AEB, 1973)

2. The reactivity of the amino ($-NH_2$) group is largely dependent on the nature of the group to which it is attached.
   Give examples which illustrate the difference in reactivity of the amino group when attached to (*a*) an alkyl group; (*b*) an aromatic nucleus; (*c*) a carbonyl group. How may this difference in reactivity by explained?
   (London, 1974)

3. (*a*) Describe how the following pairs of compounds react. In **each** case state the type of reaction taking place, give an equation and name the products of the reaction: (i) acetaldehyde and hydrogen cyanide, (ii) ethyl bromide and potassium cyanide.
   (*b*) By means of equations and brief explanatory notes suggest **one** method in **each** case by which propionitrile (ethyl cyanide) could be converted to: (i) propionic acid, (ii) n-propylamine.
   (*c*) Describe in detail how the elements carbon, hydrogen and nitrogen could be shown to be present in n-propylamine, b.p. 49 °C.
   (JMB, 1972)

Chapter 13

Some Molecules of Importance in Nature

Amino-acids

Amino-acids are the 'building bricks' of proteins, and some aspects of their chemistry therefore need to be studied. The amino-acids which are formed when proteins are broken down by hydrolysis are all α-amino-acids, in which the amino-group is attached to the α-carbon atom (the carbon atom adjacent to the carboxyl group). A number of them occur in proteins, and the simplest members of the series are the following.

$$CH_2(NH_2)CO_2H \qquad CH_3\overset{*}{C}H(NH_2)CO_2H$$

aminoacetic acid (aminoethanoic acid) (glycine) — 2-aminopropanoic acid (alanine)

Most amino-acids have trivial names by which they are usually known. In alanine, the carbon atom marked with an asterisk is asymmetric, and so alanine (and higher homologues) display optical isomerism and exist in (+)-, (−)- and racemic forms. It is usually one of the active forms which occurs naturally.

Amino-acids can be synthesized in several ways, the two most important of which are given here.

They may be obtained by treating α-halogeno-acids with ammonia:

$$\underset{\text{acetic acid (ethanoic acid)}}{CH_3CO_2H} \xrightarrow{Cl_2} \underset{\text{chloroacetic acid (chloroethanoic acid)}}{CH_2Cl.CO_2H} \xrightarrow{NH_3} \underset{\text{glycine}}{CH_2(NH_2)CO_2H}$$

Alternatively, alanine and higher homologues can be prepared from aldehydes via their cyanohydrins, by the so-called *Strecker synthesis*, e.g.

$$\underset{\text{ethanal (acetaldehyde)}}{CH_3CHO} \xrightarrow{HCN} CH_3CH(OH)CN \xrightarrow{NH_3} CH_3CH(NH_2)CN$$

$$\xrightarrow{\text{hydrolysis}} CH_3CH(NH_2)CO_2H$$

In practice the reaction is carried out in two stages as follows.

$$CH_3CHO \xrightarrow[KCN]{NH_4Cl} CH_3CH(NH_2)CN \xrightarrow[H_2O]{HCl} \underset{\text{alanine}}{CH_3CH(NH_2)CO_2H}$$

These syntheses from optically inactive starting materials give, of course, the racemic forms of the amino-acids.

Since they contain amino- and carboxyl groups, amino-acids display many of the reactions of both functions. However, the two groups influence one another's properties. In particular, since one group is basic and the other acidic, they exist as internal salts, or '*zwitterions*':

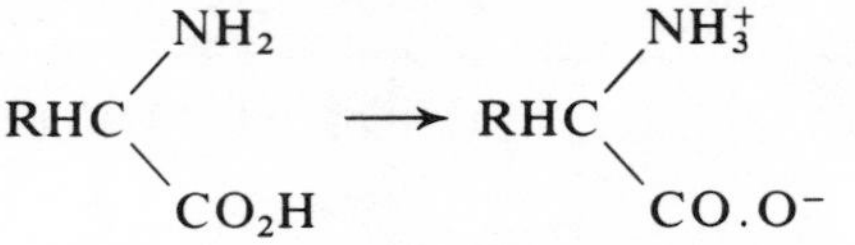

$$RHC(NH_2)(CO_2H) \longrightarrow RHC(NH_3^+)(CO.O^-)$$

Thus their physical properties are salt-like. For example, they are neither appreciably acidic nor basic, but almost neutral; they are high-melting; and they tend to be soluble in water rather than in organic solvents. For each amino-acid there is a definite pH at which, if a solution of the amino-acid is electrolysed, it does not move towards either electrode. This pH is known as the *isoelectric point* of the amino-acid.

Polypeptides

Amide-like derivatives can be formed by reaction of the amino-group of one amino-acid with the carboxyl group of another:

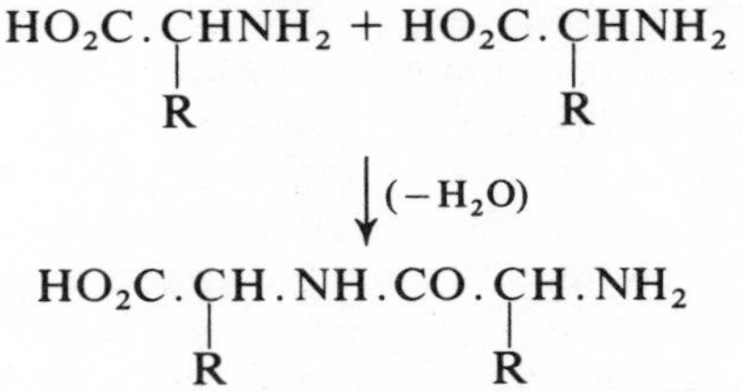

$$HO_2C.CH(R)NH_2 + HO_2C.CH(R)NH_2 \xrightarrow{(-H_2O)} HO_2C.CH(R).NH.CO.CH(R).NH_2$$

a dipeptide

The product of one such reaction which contains two amino-acid groups linked together by one '*peptide link*' is a '*dipeptide*'. The chain is obviously capable of extension at both ends, giving tripeptides, tetrapeptides, etc. The general term *polypeptides* is used for these compounds. They can be built up, stage by stage, by the use of special synthetic methods. (The simple reaction above is not generally very successful.) The point of such endeavours is to attempt to simulate natural proteins.

Proteins

Proteins are naturally occurring polypeptides with very long chains. They occur in all living cells, and a particular protein is peculiar to each type of cell. Organisms continually use up proteins (for example in gland secretions) and must replace them. Plants can synthesize them from inorganic nitrogen and other organic materials like carbohydrates, fats, etc., but animals cannot do this, and must consume proteins as an essential part of their diet to provide amino-acids from which to build up their own required proteins.

Some proteins are linear molecules, and are known as '*fibrillar*' proteins. The molecular chain lies parallel to the fibre axis. Such proteins occur in animal muscle, tendons, and hair (keratin). Others do not form fibres, and are called '*globular*' proteins. They are non-linear because of various kinds of cross-linking (cf. below). They are sometimes soluble (e.g. egg albumen). Enzymes are proteins, as also are some hormones (e.g. insulin).

The essential structural feature of proteins is that they are made up of α-amino-acid molecules joined together through peptide links.

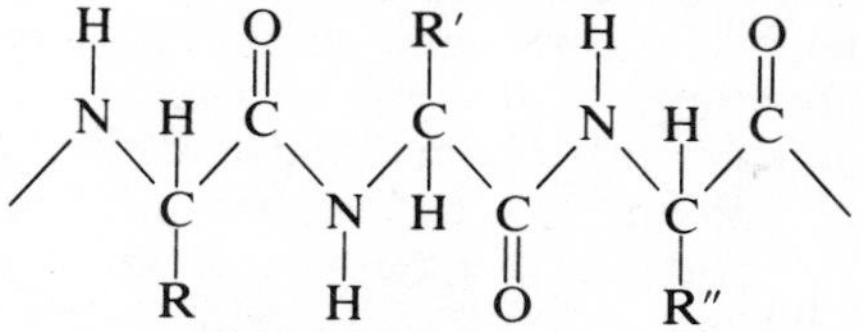

```
     H     O     R'    H     O
     |     ||    |     |     ||
     N  H  C     C     N  H  C
   /  \ | /  \ / | \  /  \ | /  \
        C     N  H  C      C
        |     |     ||     |
        R     H     O      R''
```

The groups R, R′, R″ may contain other functional groups (e.g. –OH, –SH, other amino- or carboxyl groups) which can take part in cross-links between chains leading to highly complex structures. A number of amino-acids (about twenty) occur in proteins. Some of the simpler ones apart from glycine and alanine are:

valine	$(CH_3)_2CHCH(NH_2)CO_2H$
leucine	$(CH_3)_2CHCH_2CH(NH_2)CO_2H$
serine	$CH_2(OH)CH(NH_2)CO_2H$
cysteine	$CH_2(SH)CH(NH_2)CO_2H$

Although they may be (+)- or (−)- rotatory [because there is no simple relationship between configuration and optical rotation (cf. page 155)], all of the amino-acids obtained from naturally occurring proteins belong to the same configurational family (the L-family). Thus they are all configurationally related to L-(+)-alanine and L-(−)-glyceraldehyde. It is sometimes said, for this reason, that nature is essentially left-handed.

Many amino-acids can be present in any given protein, and the *order* in which they are linked together is also characteristic. A virtually infinite number of protein structures is therefore possible in principle and, of course, the maintenance of life requires this wide variety.

In addition, many proteins occur in nature in combination with other materials: e.g. in *nucleoproteins* the proteins are associated with nucleic acids (see pages 161–9).

All proteins are hydrolysed by dilute mineral acids, and this hydrolysis ultimately gives mixtures of amino-acids.

$$\text{—NHCO—} \xrightarrow[\text{acid}]{\text{dilute}} \text{—NH}_2 + \text{HO}_2\text{C—}$$

Various polypeptides can also be formed as intermediates in this process by partial breaking up of the protein molecules.

The sizes of these giant protein molecules ('macromolecules') have been investigated by relative molecular mass measurement using various physical and chemical techniques. Some of the more reliable results have been obtained by the use of the *ultracentrifuge*, a method developed by Svedberg. The method consists essentially of measuring the rate of sedimentation of colloidal suspensions of proteins when spun at extremely high speeds in the ultra-centrifuge. Widely different values are obtained for various proteins. Insulin (R.M.M. 5733) and egg albumen (R.M.M. ~ 40 000) are relatively simple, but relative molecular masses of the order of 5×10^6 have also been measured.

The amino-acid sequence in proteins is called the *primary structure*. The chains, however, have a definite shape, or conformation; this is often, but not always, a spiral or helix. This conformation of the chains, which is stabilised largely by hydrogen-bonding, is called the *secondary structure* of the protein. In addition, the chains are often coiled and folded in complex and highly specific ways. This coiling and folding is called the *tertiary structure* of the protein. The folds are held together by various kinds of linkages between groups in different amino-acid units, often quite widely separated along the chain. These linkages are sometimes formed by salt-formation between a free amino-group on one unit and a free carboxyl group in another (which would be derived from a dicarboxylic acid, e.g. aspartic acid).

$$HO_2C.CH_2CH(NH_2)CO_2H$$

aspartic acid

$$-NH_2 + HO_2C- \longrightarrow -NH_3^+O_2C-$$

Hydrogen bonds, for example between hydroxyl and carboxyl groups, also play a part, as do disulphide linkages formed by oxidation of –SH groups which occur in cysteine.

$$-SH + HS- \underset{\text{reduction}}{\overset{\text{oxidation}}{\rightleftharpoons}} -S-S-$$

Some proteins (e.g. haemoglobin) contain several (in this case four) separate polypeptide chains which are held together in specific ways by interactions of various kinds between groups occurring in the separate chains. This final degree of complexity is called the *quaternary structure* of the protein.

Carbohydrates

Carbohydrates are among the more abundant constituents of plants and animals, and include the sugars as well as complex macromolecular materials like starch and cellulose. They serve various functions: thus sugars are sources of energy, starch acts as an energy-store, and cellulose is a main constituent of the supporting tissues of plants (e.g. wood).

Plants are able to build up carbohydrates from carbon dioxide and water by photosynthesis, but animals have to rely on plants for their supply in their diet.

Glucose and some other sugars have the empirical formula CH_2O, hence the name carbohydrate. The majority can be expressed as $C_x(H_2O)_y$, but there are some for which the ratio H:O is not 2:1. These, however, are also classed as carbohydrates because of their chemical properties.

Carbohydrates fall into two main groups, *sugars* and *non-sugars*. Sugars are generally sweet-tasting, crystalline, soluble in water, and have exact relative molecular masses. Non-sugars often have the typically colloidal properties of macromolecular substances, and relative molecular masses which are imprecise since they are averages of the masses of molecules of different sizes.

Sugars can be divided into *mono-*, *di-*, *tri-saccharides*, etc., [in general, *oligosaccharides* (*oligo-* means 'few')]. Monosaccharides cannot be hydrolysed to simpler sugars. Disaccharides can be hydrolysed with dilute acid, each molecule giving two monosaccharide molecules; e.g.

$$\underset{\substack{\text{maltose}\\\text{(a disaccharide)}}}{C_{12}H_{22}O_{11}} + H_2O \xrightarrow[\text{acid}]{\text{dilute}} \underset{\substack{\text{glucose}\\\text{(a monosaccharide)}}}{2C_6H_{12}O_6}$$

Thus a disaccharide consists of two monosaccharide units joined together with elimination of one molecule of water. Trisaccharides contain three monosaccharide units, tetrasaccharides four, etc. The most common sugars are mono- and di- saccharides.

The non-sugars are *polysaccharides* since they are macromolecular structures built up similarly from many (n) monosaccharide units joined together with the loss of ($n - 1$) molecules of water.

Sugars

Monosaccharides can be classified according to the number of carbon atoms they contain. The commonest are *hexoses*, which have the molecular formula $C_6H_{12}O_6$. Glucose and fructose are both hexoses with this molecular formula, and are therefore isomeric. *Pentoses* are also common. Ribose, which is a constituent of ribonucleic acid (RNA), is a pentose, and a reduction product of ribose (deoxyribose) is the corresponding constituent of deoxyribonucleic acid (DNA).

Some sugars give the reactions of aldehydes, and are called *aldoses*, others behave like *ketones* and are called *ketoses*. For example, glucose is an aldose, or more specifically, an *aldohexose*, while fructose is a *ketohexose*.

Glucose (or 'dextrose') is found in most fruits, and is also formed by the complete hydrolysis of starch or cellulose. Since it contains asymmetric carbon atoms (see below) it exists in (+)- and (−)- forms which are non-superposable mirror images of one another. The common, naturally occurring form is D(+)-glucose. The prefix (+)- indicates that it rotates the plane of polarization to the right (i.e. clockwise), and it is said to be *dextrorotatory*. The prefix D specifies the configuration of the molecule (i.e. which enantiomer it is). *There is no simple relationship between configuration and the sign of the rotation.* Thus, for example, the common enantiomer of fructose is also the D form, that is, it is in the same configurational family as D(+)-glucose, and not in the opposite 'mirror image' (L) family. However, D-fructose is strongly laevorotatory (it rotates the plane of polarization to the left) and is therefore D(−)-fructose.

Glucose is typical of aldoses and behaves as a strong reducing agent. On oxidation it gives a carboxylic acid with the same number of carbon atoms as itself. It also reduces Fehling's solution and Tollen's reagent (Chapter 10, page 105), and thus behaves as an aldehyde. (Sugars which exhibit these reactions are known as '*reducing sugars*'.)

It reacts with acetic anhydride, giving a penta-acetate, and therefore contains five hydroxyl groups. Other chemical evidence indicates that the chain is unbranched. All this evidence indicates that the structure of glucose might be written as:

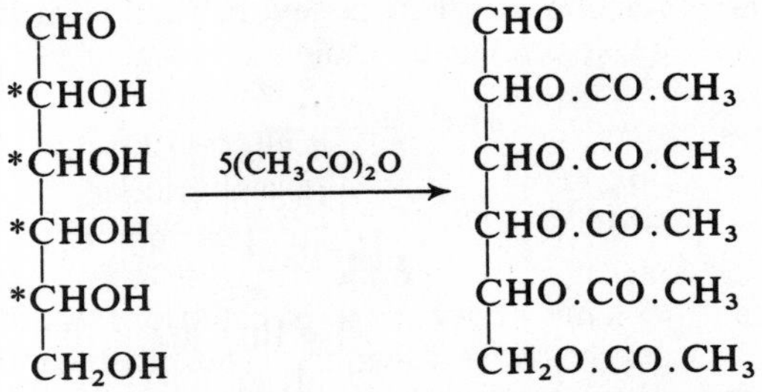

$$\begin{array}{l} CHO \\ |\\ {}^{*}CHOH \\ | \\ {}^{*}CHOH \\ | \\ {}^{*}CHOH \\ | \\ {}^{*}CHOH \\ | \\ CH_2OH \end{array} \xrightarrow{5(CH_3CO)_2O} \begin{array}{l} CHO \\ | \\ CHO.CO.CH_3 \\ | \\ CHO.CO.CH_3 \\ | \\ CHO.CO.CH_3 \\ | \\ CHO.CO.CH_3 \\ | \\ CH_2O.CO.CH_3 \end{array}$$

(the penta-acetate)

The four asterisked carbon atoms in this structure are asymmetric, and all different from one another. This leads to the existence of sixteen stereoisomers, which consist of eight pairs of mirror images. There are in fact eight aldohexoses, each with D and L enantiomers, making sixteen in all. The eight are called allose, altrose, glucose, mannose, gulose, idose, galactose, and talose. Thus, for example, there is an L(−)-glucose whose structure is the mirror image of D(+)-glucose and which has an equal and opposite optical rotation.

Further chemical evidence indicates, however, that these 'open-chain' structures are not quite correct. For example, if they were, and glucose therefore had an aldehyde group at the end of a chain, it should react with *two molecules* of an alcohol (e.g. methanol, MeOH) to give an acetal (cf. Chapter 10, page 107).

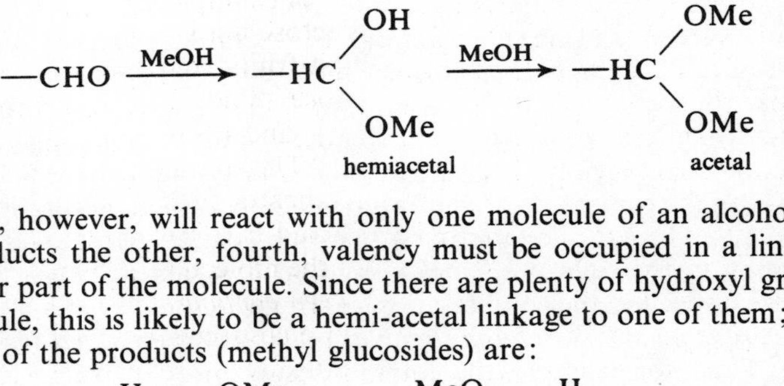

$$—CHO \xrightarrow{MeOH} —HC\begin{matrix} OH \\ OMe \end{matrix} \xrightarrow{MeOH} —HC\begin{matrix} OMe \\ OMe \end{matrix}$$

hemiacetal — acetal

Aldoses, however, will react with only one molecule of an alcohol, so in these products the other, fourth, valency must be occupied in a linkage to some other part of the molecule. Since there are plenty of hydroxyl groups in the molecule, this is likely to be a hemi-acetal linkage to one of them; i.e. the structures of the products (methyl glucosides) are:

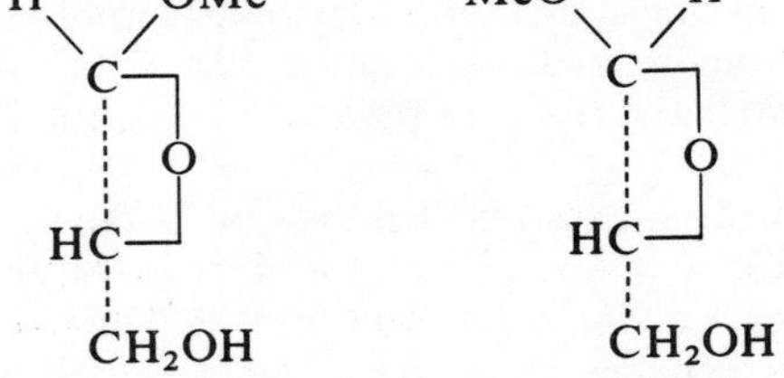

$$\begin{array}{c} H \quad OMe \\ \diagdown \; \diagup \\ C— \\ \vdots \quad | \\ \vdots \quad O \\ \vdots \quad | \\ HC— \\ \vdots \\ CH_2OH \end{array} \qquad \begin{array}{c} MeO \quad H \\ \diagdown \; \diagup \\ C— \\ \vdots \quad | \\ \vdots \quad O \\ \vdots \quad | \\ HC— \\ \vdots \\ CH_2OH \end{array}$$

α– β–

There are two isomeric forms, known as the α- and β- methyl glucosides, as shown above.

While this evidence does not tell us anything about the structures of sugars themselves, but only of the methyl derivatives, it does strongly suggest that the sugars have analogous structures, i.e. that they are already internal hemi-acetals in which the aldehyde group takes part in a hemi-acetal linkage with one of the hydroxyl groups. Other evidence which need not be described in detail here shows that this is so, and also that the hydroxyl group involved is that on the fifth carbon atom (C5) giving a six-membered ring, although in certain circumstances and in certain sugars, C4 can be involved, giving a five-membered ring. Now, as in the methyl glucosides, two forms (α- and β-) should exist, because the hemiacetal formation renders C1 asymmetric, whereas it is not so in the open chain ('aldehydic') structure. Thus there should be two forms of D-glucose as shown in the formulae below. (In these formulae the configurations of C2, C3, C4, and C5 are specified, so that the formulae shall be correct. These configurations have been established for all the sugars by experiment and reasoning).

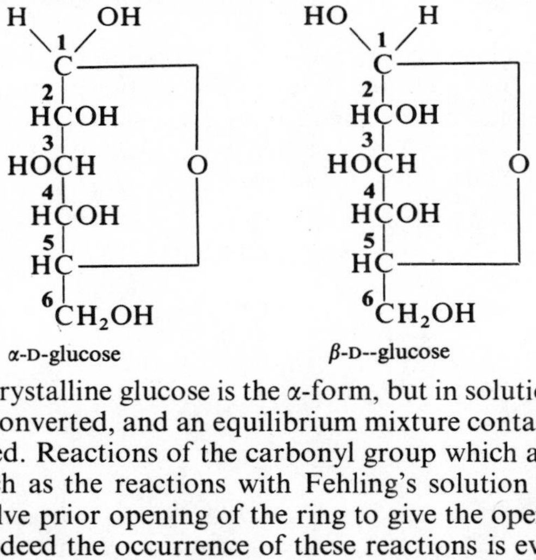

H OH
C1
HC2OH
HOC3H O
HC4OH
HC5
C6H_2OH

α-D-glucose

HO H
C1
HC2OH
HOC3H O
HC4OH
HC5
C6H_2OH

β-D--glucose

The common crystalline glucose is the α-form, but in solution the two forms are rapidly interconverted, and an equilibrium mixture containing both forms is soon established. Reactions of the carbonyl group which are undergone by these sugars (such as the reactions with Fehling's solution and Tollen's re-agent) must involve prior opening of the ring to give the open-chain forms as intermediates. Indeed the occurrence of these reactions is evidence that very small amounts of the open-chain forms exist in solution in equilibrium with the cyclic forms.

Although formulae written like this serve to illustrate the relationship between the open-chain and cyclic forms, they give an entirely wrong impres-sion of the shapes of sugar molecules. Formulae written as follows are much better.

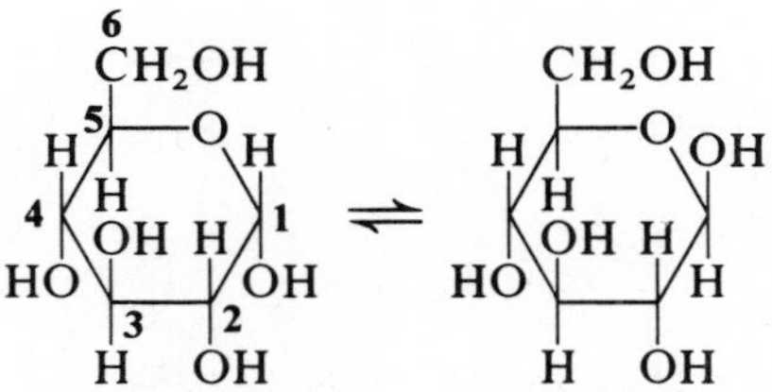

6
CH_2OH
H 5 O H
4 H 1
OH H
HO OH
3 2
H OH

α-D-glucose

⇌

CH_2OH
H O OH
H
OH H
HO H
H OH

β-D-glucose

Even these formulae are not entirely satisfactory because the rings, like those of cycloalkanes (Chapter 3), are not planar owing to the tetrahedral disposition of valencies about carbon. Although the above formulae are generally used for simplicity and convenience, it is sometimes necessary to show the actual shapes of the molecules more accurately, and then formulae such as those below are used.

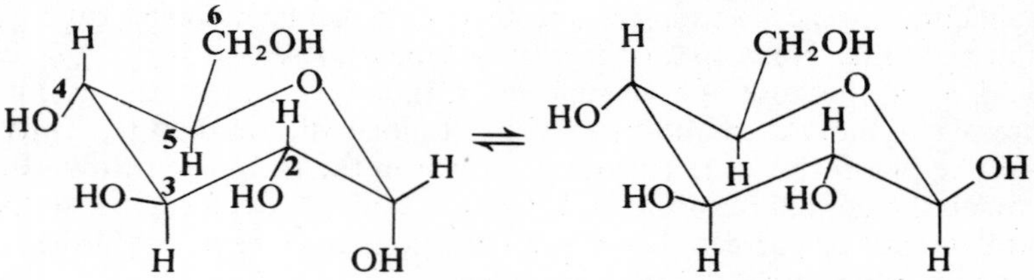

α-D-glucose ⇌ β-D-glucose

Fructose, which also occurs in fruits, is known to behave as a ketone by its reactions, many of which are those of the carbonyl group. Although it is quite readily oxidized it gives mixtures of acids with fewer carbon atoms than itself, and it is therefore ketonic rather than aldehydic. The keto-group is on C2, and its open chain form is therefore:

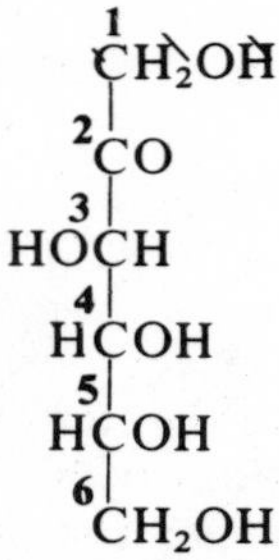

$$
\begin{array}{c}
{}^{1}CH_2OH \\
| \\
{}^{2}CO \\
| \\
HO\overset{3}{C}H \\
| \\
H\overset{4}{C}OH \\
| \\
H\overset{5}{C}OH \\
| \\
{}^{6}CH_2OH
\end{array}
$$

D(−)-fructose

It is worth noting that the configurations of the –H and –OH groups on C3, C4, and C5 of fructose are the same as they are in glucose, and glucose and fructose are therefore close structural relations.

Once again the reactions of fructose have shown that it is an internal hemiacetal, and its ring structure is as follows:

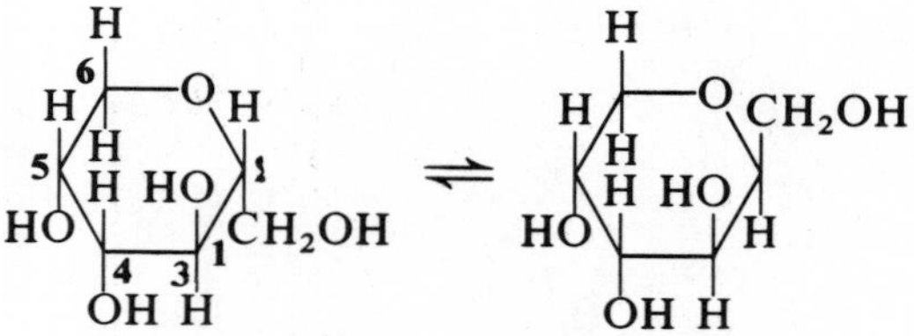

α- and β-D-fructose

Disaccharides

The linkage between the monosaccharide units in disaccharides is known as the *glycosidic linkage.* It is an oxygen-bridge, as in acetals such as the methyl glucosides. The disaccharide *maltose* which is present in malt ($C_{12}H_{22}O_{11}$) contains two glucose units in which C1 of one is linked to C4 of the other by an α-glycosidic linkage (i.e. its configuration is as in α- rather than β-methyl glucoside). Maltose can be hydrolysed to glucose either by heating with acid, or enzymically, by the enzyme maltase which is present in yeast. Yeast also contains another enzyme *zymase,* which is responsible for the conversion of glucose into ethanol in the fermentation process (page 90).

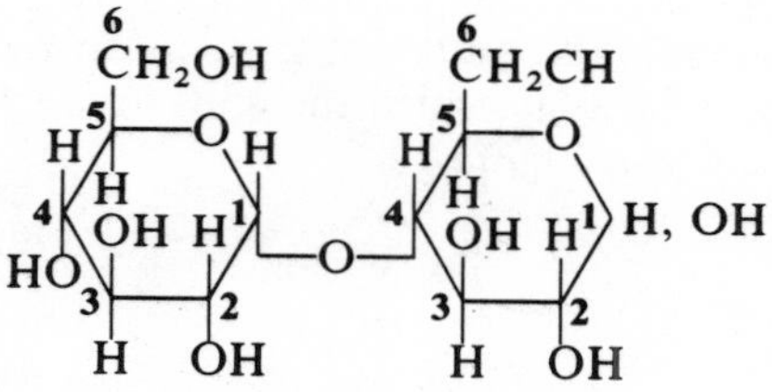

maltose

The C1 of the right hand glucose unit can have α- and β- configurations, so it is left unspecified, as shown above. Moreover, this is the aldehyde group when in the open chain form, i.e. the 'reducing group'. It is not involved in a glycosidic linkage, and is hence free to take part in aldehyde reactions in maltose. Maltose is therefore a reducing sugar, i.e. it reduces Fehling's solution and Tollen's reagent.

Another very important disaccharide is *sucrose* (also $C_{12}H_{22}O_{11}$, cane sugar, beet sugar, ordinary table sugar). Structurally sucrose is rather unusual. The component sugars are glucose and fructose, and the C1 of glucose is glycosidically linked to the C2 of fructose. Thus both reducing groups are involved in the linkage and neither is free to display its characteristic reactions. Sucrose is therefore a non-reducing sugar, i.e. it does not react with Fehling's solution or Tollen's reagent.

Moreover, the fructose unit in sucrose is not in the ordinary six-membered ring form as in fructose itself, but instead in a five-membered ring ('*furanose*') form. The structure of sucrose is:

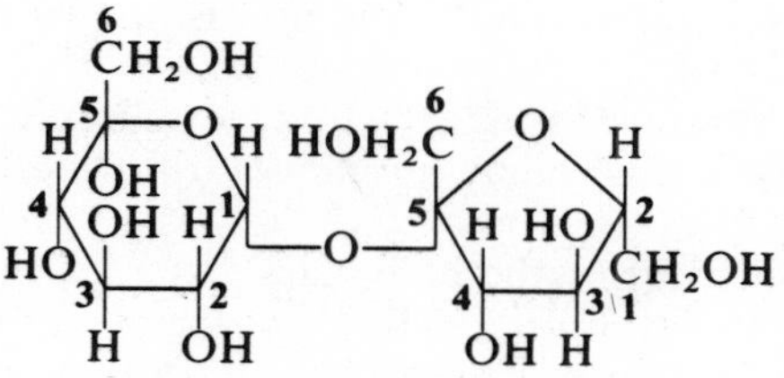

(glucose) (fructose)

sucrose

Polysaccharides

The two most important polysaccharides are starch and cellulose, and both consist of glucose units.

Cellulose is the chief constituent of the cell walls of plants. Cotton wool is almost pure cellulose. It is insoluble in all the usual solvents, but will dissolve in Schweitzer's reagent (ammoniacal copper hydroxide).

It can be hydrolysed with acids to give, ultimately, glucose. Some esters of cellulose are used as artificial fibres (e.g. acetates in rayons of various kinds). The trinitrate 'nitrocellulose' is guncotton, an explosive.

The glycosidic linkage in cellulose is between C1 of one unit and C4 of the next. Its configuration is β, and this results in the units being arranged in a 'zig-zag' way, so that the molecules do not coil up. This is why cellulose forms threads, since these consist of bundles of these linear molecules held together by hydrogen bonds. The way in which the units are arranged in the chains is shown diagrammatically in Figure 13.1.

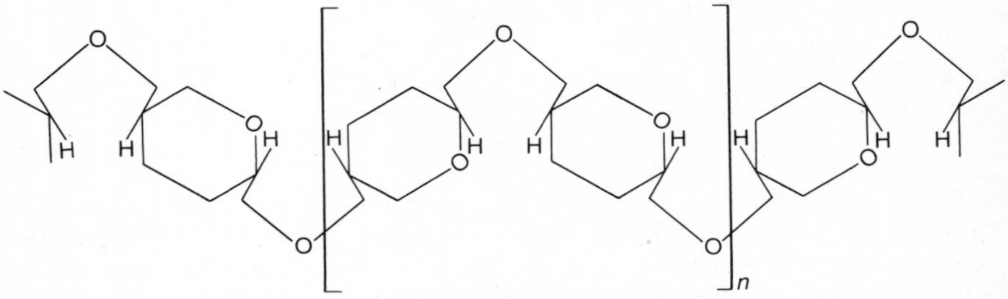

Figure 13.1 Cellulose (diagrammatical)

Relative molecular mass measurements by physical and chemical methods (including the use of the ultracentrifuge) have given values ranging from 20 000 to 500 000, corresponding to chain lengths of 150 to 3500 glucose units.

Starch is the carbohydrate reserve material of plants. It is an amorphous solid, which gives colloidal solutions in water. Again it can be hydrolysed ultimately to glucose and therefore consists of glucose units. Starch is, however, not homogenous, but can be separated into two components, *amylose* and *amylopectin*.

Amylose is the simpler of these, and is like cellulose, in that it consists of an unbranched chain of glucose units linked 1,4; but with the difference that the linkage is now α-, rather than β-, in configuration. The molecule is therefore not a zig-zag as in cellulose, but coils up, as shown diagrammatically in Figure 13.2.

The long chains therefore assume the form of helices, with about six glucose units in each turn. This is why amylose cannot form good straight threads like cellulose.

It is amylose which forms the blue-black complex of starch with iodine. In this complex the iodine molecules lie along the axis of the helix.

α-Glycosidic linkages are hydrolyzed by the enzyme maltase. Thus starch gives the disaccharide maltose and this is what occurs in the malting process when barley is allowed to germinate in a warm moist environment. It is also the first essential process in our assimilation of the starch we eat. We cannot assimilate cellulose in the same way, because we cannot break the β-glycosidic linkages. This requires another enzyme, called emulsin, which we do not have. Many other animals (e.g. cattle) do have emulsin, however, and this is why they can feed on grass while we cannot.

The chains in amylose contain anything between 60 and 6000 glucose units.

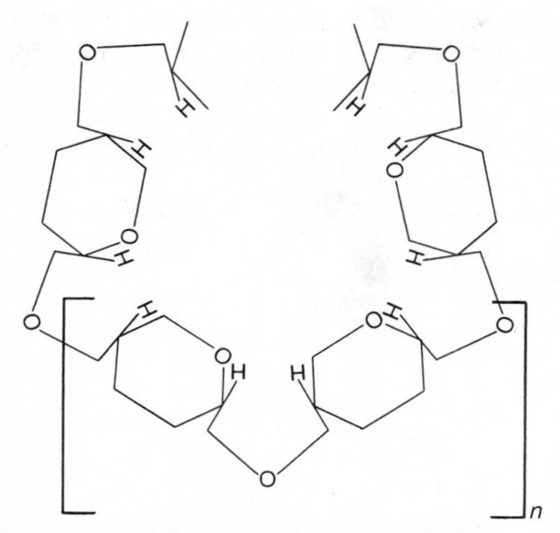

Figure 13.2 Amylose (diagrammatical)

Amylopectin, the other component of starch, also contains glucose units linked α-1,4, but there are also many chain branches involving 1,6 and other linkages. It is thus a highly complex structure, and relative molecular masses of 50 000 to 10 000 000 have been recorded for it (i.e. 350–70 000 glucose units).

Nucleic acids

Nucleic acids are macromolecular substances which occur in all living cells. Their importance is that the way in which a nucleic acid molecule is built up itself constitutes a set of instructions for the metabolic activities of the cell. Specifically, nucleic acids govern the order in which the cell 'strings together' amino-acid molecules to synthesize proteins.

Nucleic acids can be broken down by hydrolysis, first to the units of which they are composed, which are called *nucleotides*. Nucleotides can be further broken down to simpler substances called *nucleosides* and phosphoric acid. Nucleotides are therefore phosphate esters of nucleotides. Each nucleoside can itself be finally broken down into two components, namely a sugar (ribose in ribonucleic acid, RNA; or deoxyribose in deoxyribonucleic acid, DNA, both sugars being in their furanose forms), and one of several organic bases.

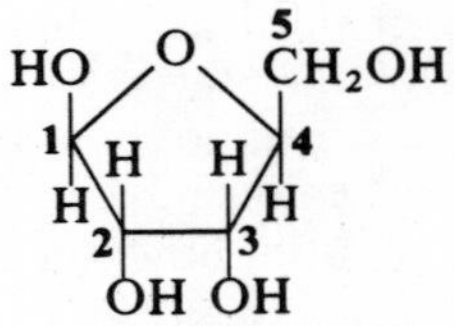

ribose (furanose form)

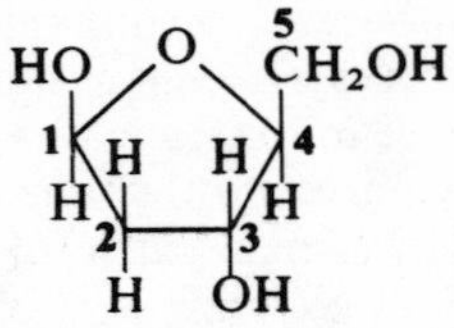

deoxyribose (furanose form)

The bases concerned are *uracil*, *thymine*, and *cytosine*, which are derivatives of the parent substance *pyrimidine*, and are called the pyrimidine bases; and *adenine* and *guanine*, which are derivatives of *purine*, and are called the purine bases.

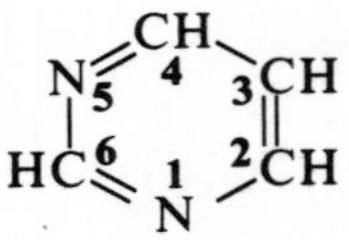

pyrimidine

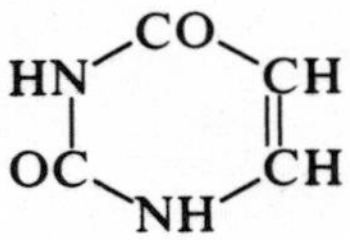

uracil

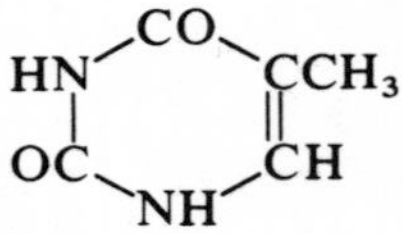

thymine

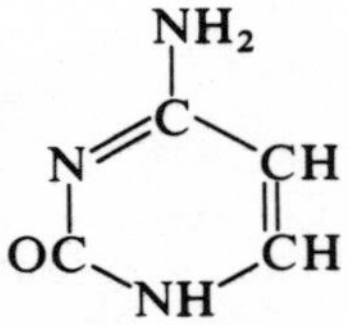

cytosine

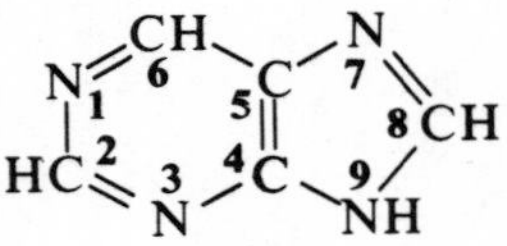

purine

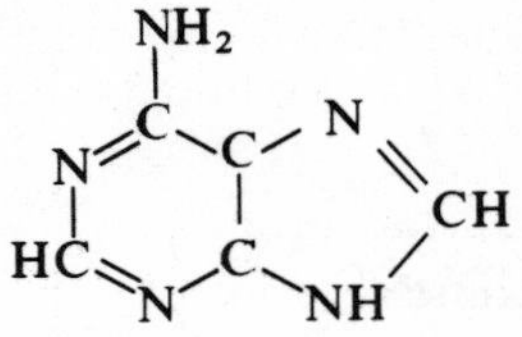

adenine

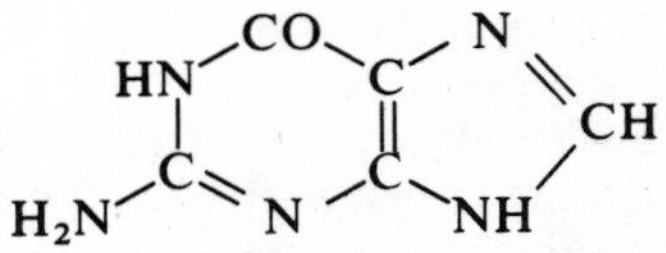

guanine

In the pyrimidine nucleosides the bond to the sugar is from position 1 of the base to C1 of the sugar, e.g.

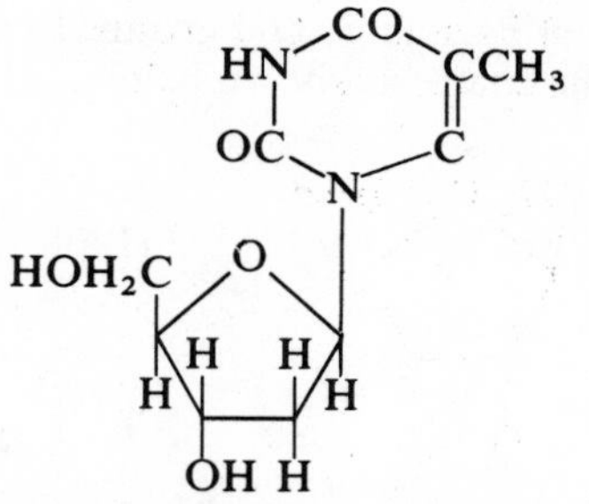

deoxythymidine (occurring in DNA)

In the purine nucleosides, position 9 of the base and C1 of the sugar are involved, e.g.

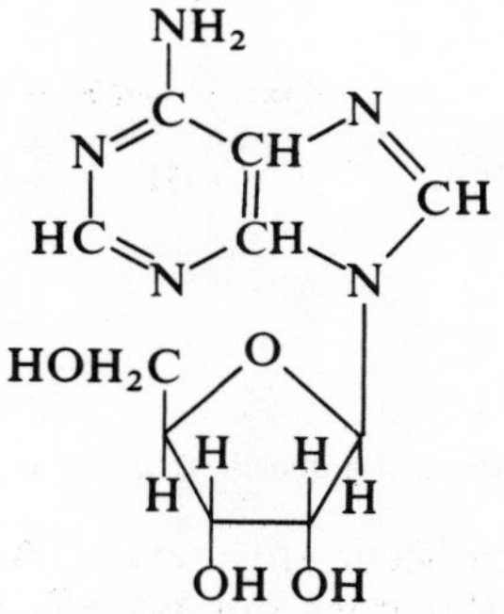

adenosine (occurring in RNA)

In the nucleotides, the phosphate group is attached to C5 of the sugar, e.g.

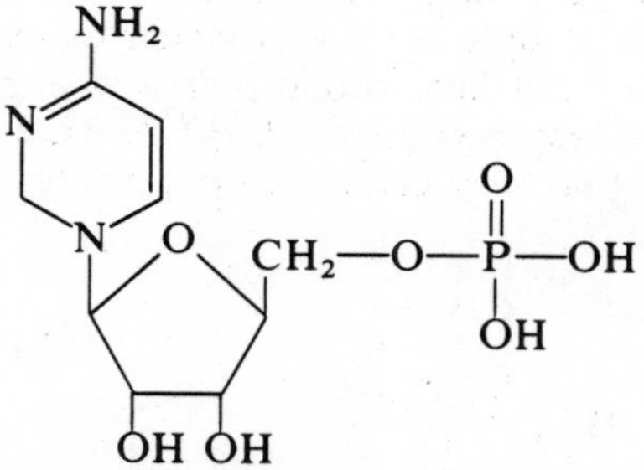

adenylic acid (adenosine monophosphate)

(In the above formula, as in conventional representations of benzene derivatives, carbon and hydrogen atoms in rings are omitted for simplicity).

Nucleic acids are polymers of nucleotides in which the phosphate group is linked through another of its acidic –OH groups to position 3 of the sugar of the next nucleotide in the chain, as shown below.

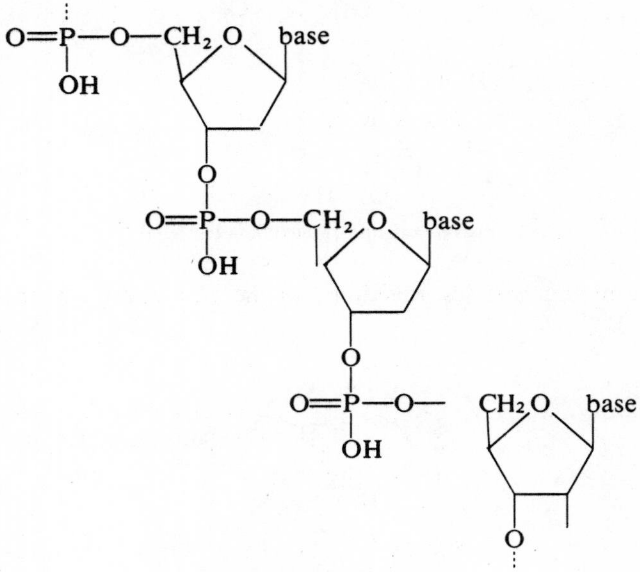

part of a nucleic acid chain

In DNA, the bases which occur are adenine (A), guanine (G), cytosine (C), and thymine (T) [uracil (U) does not occur]. In RNA adenine, guanine, cytosine, and uracil occur, but thymine does not.

DNA–the double helix

DNA consists of two polynucleotide chains coiled together so as to form a double helix. The two strands are held together by hydrogen bonds between bases, but the crucial point here is that *only certain pairs* of bases can form strong linkages in this way because the C=O, N–H, and =N groups involved must be arranged so that they come into proximity. Thus thymine can pair with adenine:

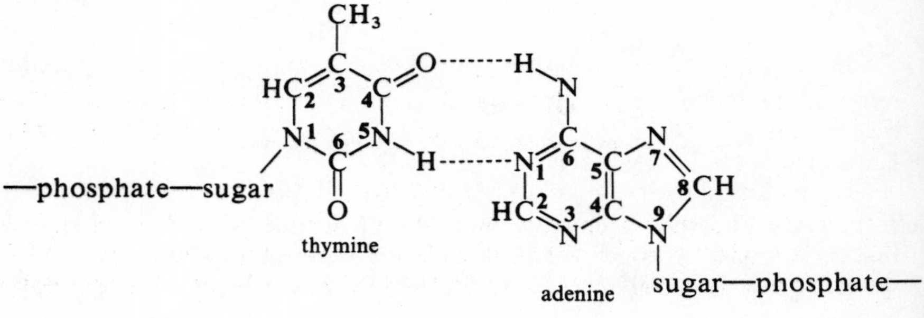

—phosphate—sugar   thymine   adenine   sugar—phosphate—

A careful examination of the structures of the bases (cf. page 162) will show that two pairs only are possible in DNA, namely adenine–thymine (as above) and cytosine–guanine (see below), although the order of the members of each pair can be reversed, so that there are four combinations altogether.

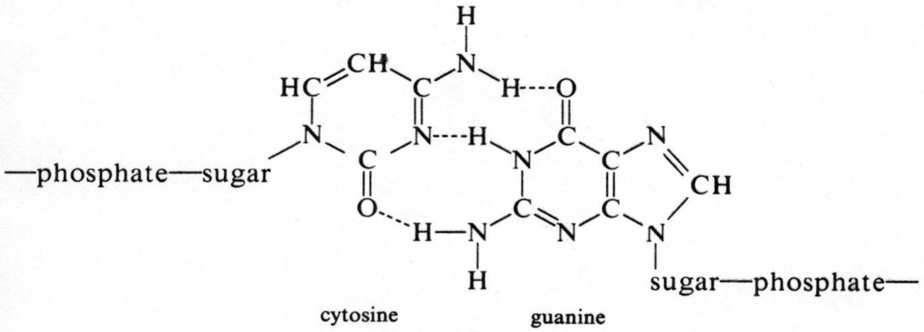

—phosphate—sugar

sugar—phosphate—

cytosine guanine

The four combinations, and the way in which the two chains (which are 'upside down' with respect to one another) are linked, are shown in Figure 13.3.

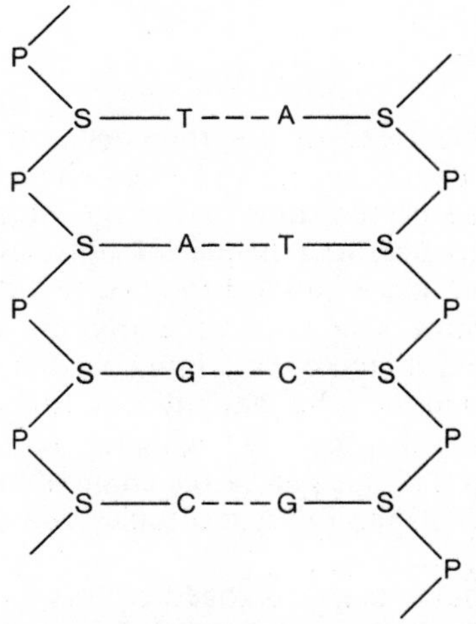

S—T---A—S

S—A---T—S

S—G---C—S

S—C---G—S

Figure 13.3 Arrangement of phosphate groups (P), sugar (deoxy-ribose, S), and bases (T, A, G, and C) in part of a DNA molecule

When arranged in the double helix, the two chains appear as in Figure 13.4, where linkages between chains through bases and hydrogen bonds are shown by broken lines. The bases project inwards towards the centres of the helices. Each turn of the helices is about 3.4 nm long.

This picture enables us to understand how DNA can *replicate*, or reproduce itself, so that in a cell, in the presence of enzymes which catalyze the linking together of nucleotides through phosphate groups, two *identical* daughter molecules can be formed from one parent molecule. In both daughter

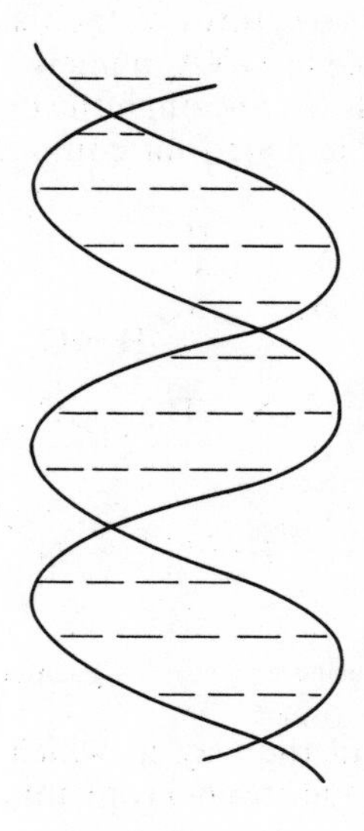

Figure 13.4

molecules, the same nucleotides occur in the same order as in the parent. What happens is that the two chains uncoil, and a new chain is built up on each. The new double chains must be the same as the old because of the restrictions mentioned on page 165 concerning the pairing of bases. Thus, for example, an adenine-containing unit, which had to be joined to a thymine-containing unit in the parent chain, must also take up a thymine-containing unit in the daughter. Likewise its old thymine-containing partner must again link with an adenine-containing unit in the other daughter. In this way the daughters must be identical both with each other and with the parent. The uncoiling and building up process starts at one end of the chain and proceeds along it. This process must happen to all the DNA molecules in a cell during mitosis (cell division).

Although the two chains are held together very firmly by the cumulative effect of all the hydrogen bonds along them, it is energetically relatively easy for them to be 'unzipped' from one end, since the dissociation energy of hydrogen bonds is only of the order of 25 kJ mol^{-1}, so that each individual hydrogen bond requires only a small amount of energy to break it. Thus the above implicit analogy of a zip-fastener is a good one: while the cumulative effect of all the links in a zip makes it impossible to tear the two halves apart, each individual link is easily parted, so that the fastener is easily opened from one end.

The 'double helix' structure and the replication mechanism were first postulated by F. H. C. Crick and J. D. Watson. Much of the experimental work which led to these postulates had previously been done by E. Chargaff and M. Wilkins, among others.

Ribonucleic acid (RNA)

RNA contains ribose instead of deoxyribose, and uracil occurs in it instead of thymine as in DNA. Moreover it has only one strand in its molecule. There are several types of RNA. One, *messenger RNA* (mRNA), is synthesized on the strands of uncoiling DNA by enzymatic stringing together of nucleotides as described above for DNA replication. A different enzyme is necessary, and this time it is a uracil-containing nucleotide which enters the chain when adenine occurs in the DNA 'template', rather than thymine, which would have entered if DNA were being replicated. Thus the mRNA molecule contains the same information as the DNA on which it was synthesized.

A second type of RNA is also necessary for protein synthesis. This actually carries the amino-acids during the protein synthesis and is called *transfer RNA* (tRNA). It has a relatively short chain (70–80 nucleotides) of which one end always terminates in the sequence of bases –C–C–A. The molecule is coiled on itself, and base-pairing, as in DNA, occurs to hold it in this form. A sequence of three bases projects at the point where the chain turns back on itself. This triplet of bases is called an *anticodon*, and its function is explained in Figure 13.5.

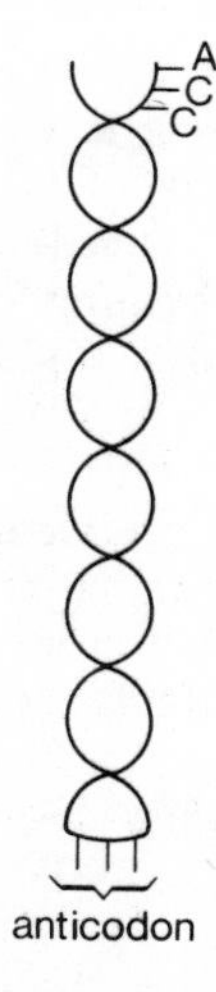

Figure 13.5 Transfer RNA

Protein synthesis : the genetic code

Proteins are synthesized at structures in the cell called *ribosomes* which consist partly of a third RNA, called ribosomal RNA (rRNA), and protein. DNA uses mRNA, which has been synthesized on it, to send in code the required sequence of amino-acids to the ribosomes. This information is contained in the sequence of bases along the mRNA chain. Since there are twenty amino-acids to be coded, and only four different bases, more than one base is needed to code for each amino-acid. In fact it is groups of three adjacent bases which are the coding units or *codons*. Usually more than one triplet of bases will code for a given amino-acid, since there are 64 different triplets for 20 amino-acids. Thus, for example, the triplets GGU, GGC, GGA, and GGG code for glycine and GCU, GCC, GCA, and GCG for alanine. It appears that the third base can vary somewhat, but the first and second do not. The code has been 'cracked'. and the triplets coding for each of the 20 amino-acids are now known.

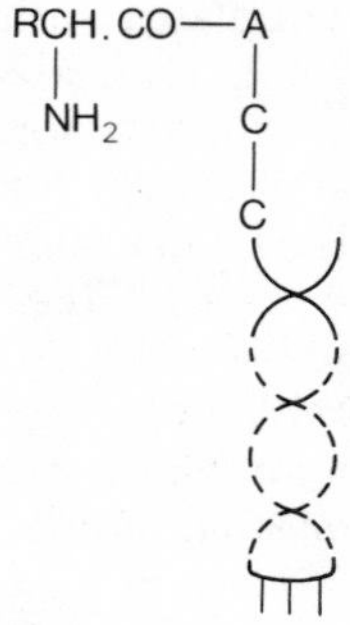

RCH.CO—A
NH_2
C
C

Figure 13.6 A charged tRNA molecule

Ribosomes attach themselves to mRNA carrying the coded instructions along its chain. The carboxyl groups of amino-acids become attached, by a sequence of enzymic processes which need not concern us here, to the adenine units at the ends of tRNA chains, forming 'charged' tRNA molecules (Figure 13.6). Each kind of tRNA molecule is highly specific for its own particular amino-acid, so that a tRNA with a particular triplet of bases as its anticodon will carry a particular amino-acid. This occurs in such a way that the anticodon can attach itself, by inter-base hydrogen bonding, to the codon on the mRNA chain which codes for the particular amino-acid carried by the tRNA molecule of which it forms a part. Thus, for example, a tRNA with a CCC anticodon will carry glycine, and will attach itself to a GGG codon on the mRNA chain, GGG being one of the triplets which codes for glycine.

Now if the first codon on the mRNA chain is, for example, GGG, a charged tRNA molecule with anticodon CCC and carrying glycine will attach itself to the chain at the ribosome (Figure 13.7).

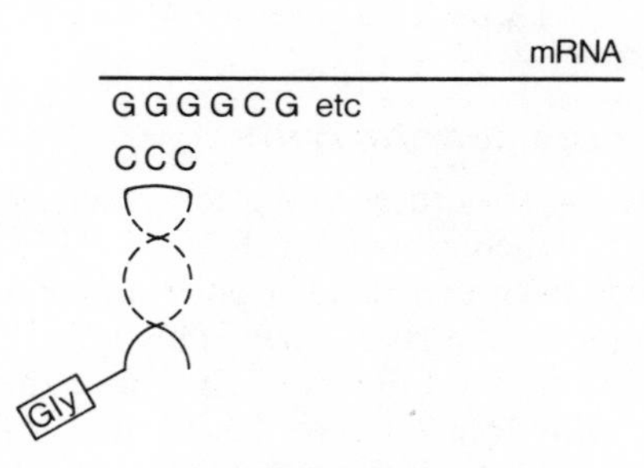

Figure 13.7

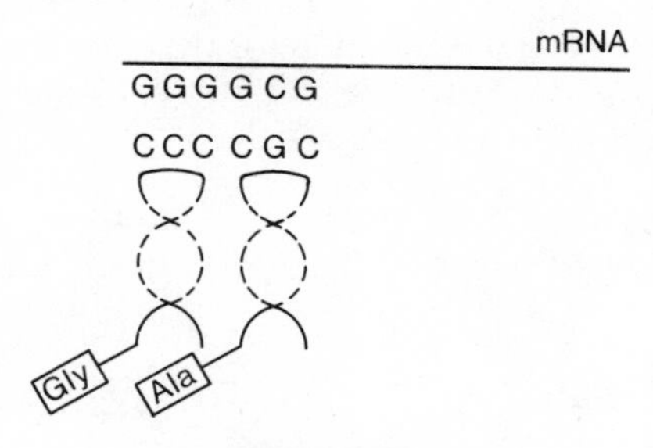

Figure 13.8

If the next codon is GCG, say, it will need a charged tRNA with the anti-codon CGC, and carrying alanine (for which GCG is a code), to attach itself (Figure 13.8). The glycine is then in the right position to attach itself to the alanine and the first tRNA drops off. The appropriate charged tRNA then attaches itself to the next mRNA triplet. As illustrated in Figure 13.9, this is CCA, which codes for the amino-acid proline. The building up of the protein chain continues in this way right along the mRNA chain until it gets to the end. Thus a polypeptide chain for the protein is built up of precisely the right length, containing the right amino-acids in the right sequence.

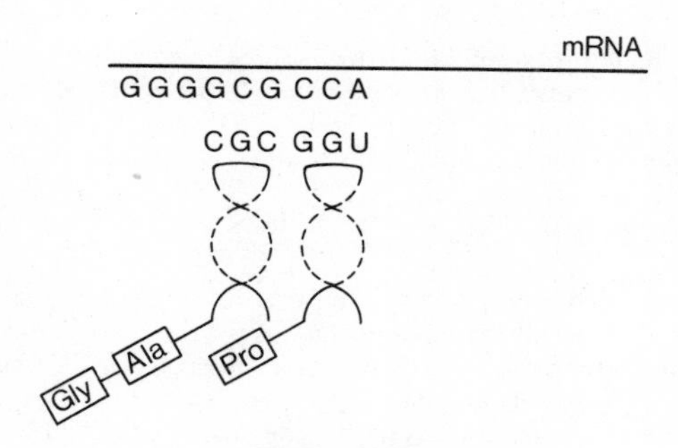

Figure 13.9

Adenosine triphosphate (ATP)

Phosphate groups are characterized by their ability to be further phosphorylated by uptake of additional phosphate groups.

$$\underset{\text{monophosphate}}{-O-\overset{\overset{\displaystyle O}{\|}}{\underset{\underset{\displaystyle OH}{|}}{P}}-OH} + \underset{\text{phosphoric(V) acid}}{HO-\overset{\overset{\displaystyle O}{\|}}{\underset{\underset{\displaystyle OH}{|}}{P}}-OH} \longrightarrow \underset{\substack{\text{diphosphate}\\ \text{[systematically,}\\ \text{heptaoxodiphosphate(V)]}}}{-O-\overset{\overset{\displaystyle O}{\|}}{\underset{\underset{\displaystyle OH}{|}}{P}}-O-\overset{\overset{\displaystyle O}{\|}}{\underset{\underset{\displaystyle OH}{|}}{P}}-OH}$$

Thus the nucleotide adenosine monophosphate can become phosphorylated twice to form adenosine triphosphate (ATP), a substance of great biochemical importance which occurs in all types of cells.

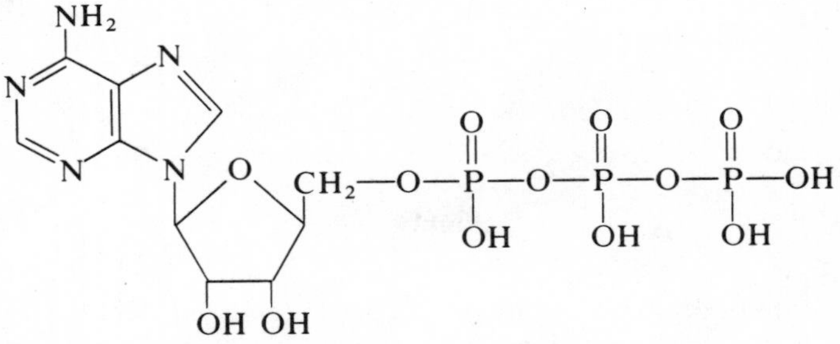

adenosine triphosphate

Free energy is released by the reverse of this process, the hydrolysis of ATP to the diphosphate (ADP). Thus organisms can use ATP as an energy-store.

Questions

1. Write a general account of the structure and properties of the main types of organic compounds found in living matter, indicating where they occur in living organisms and what their function is.

(NISEC, 1973)

2. Describe in outline the synthesis of 2-aminopropanoic acid (α-aminopropionic acid, α-alanine) from any suitable starting material. How does the synthetic material differ from that found in nature?

What is the action on this compound of (*a*) acetic anhydride, (*b*) nitrous acid, (*c*) sodium hydroxide solution?

Explain what you understand by the term 'peptide'.

(O and C Special Paper, 1972)

3. (*a*) Polymerisation reactions may be classified as addition or condensation reactions. Explain the meaning of these terms, illustrating your answer with two named examples in each case. How would you demonstrate in the laboratory one such polymerization process?

(*b*) Compare the structures of the simple monosaccharides, glucose and fructose. How do solutions of these compounds affect (i) plane polarised light, and (ii) ammoniacal solutions of silver nitrate?

Indicate by means of a block diagram the structure of a polysaccharide.

(London, 1975)

4. (*a*) Outline the preparation of pure glycine from acetic acid. How, and under what conditions, does glycine react with (i) hydrochloric acid, (ii) sodium hydroxide?

Explain why glycine is soluble in water and insoluble in ether.

(*b*) Describe **two** general methods for the preparation of esters.

(*c*) What are the essential structural features of each of the following biologically important compounds?

(i) polypeptides,
(ii) urea,
(iii) naturally occurring fats.

(JMB, 1971)

Index

Individual compounds are not generally indexed, except where mention of them is particularly helpful. Where one of several references is much more important than the rest it is given in heavy type.